Ecological Protection and the Environment

Ecological Protection and the Environment

Edited by **Andrew Hyman**

SYRAWOOD
PUBLISHING HOUSE
New York

Published by Syrawood Publishing House,
750 Third Avenue, 9th Floor,
New York, NY 10017, USA
www.syrawoodpublishinghouse.com

Ecological Protection and the Environment
Edited by Andrew Hyman

© 2016 Syrawood Publishing House

International Standard Book Number: 978-1-68286-001-4 (Hardback)

Printed in the United States of America.

Contents

Preface

Scientists and ecologists from across the globe are striving to asses and evaluate the damage caused to earth's ecosystems and environment because of human interference. This book highlights the importance of maintaining equilibrium in the human-environment interaction. Different approaches, evaluations, and advanced studies on ecological protection have been included in this book along with topics such as impact of climate change, waste management, conservation and management of water resources, etc. As this field is emerging at a rapid pace, the contents of this book will help the readers understand the modern concepts and significance of the subject.

The researches compiled throughout the book are authentic and of high quality, combining several disciplines and from very diverse regions from around the world. Drawing on the contributions of many researchers from diverse countries, the book's objective is to provide the readers with the latest achievements in the area of research. This book will surely be a source of knowledge to all interested and researching the field.

In the end, I would like to express my deep sense of gratitude to all the authors for meeting the set deadlines in completing and submitting their research chapters. I would also like to thank the publisher for the support offered to us throughout the course of the book. Finally, I extend my sincere thanks to my family for being a constant source of inspiration and encouragement.

Editor

Combining Active and Passive Airborne Remote Sensing to Quantify NO$_2$ and O$_x$ Production near Bakersfield, CA

Sunil Baidar[1,2], Rainer Volkamer[1,2,*], Raul Alvarez[3], Alan Brewer[3], Fay Davies[4], Andy Langford[3], Hilke Oetjen[1], Guy Pearson[5], Christoph Senff[2,3] and R. Michael Hardesty[2,3]

[1]Department of Chemistry and Biochemistry, University of Colorado, Boulder, CO, USA.
[2]Cooperative Institute for Research in Environmental Sciences, Boulder, CO, USA.
[3]Earth System Research Laboratory, NOAA, Boulder, CO, USA.
[4]School of Built Environment, University of Salford, Salford, UK.
[5]Halo Photonics, Worcestershire, UK.

Authors' contributions

Authors RV, CS and RMH designed research; all authors performed research. Authors SB, HO, AB and CS analyzed data; authors SB and RV interpreted data and wrote the paper. All authors provided comments.

ABSTRACT

Aims: The objective of this study is to demonstrate the integrated use of passive and active remote sensing instruments to quantify the rate of NO$_x$ emissions, and investigate the O$_x$ production rates from an urban area.
Place and Duration of Study: A research flight on June 15, 2010 was conducted over Bakersfield, CA and nearby areas with oil and natural gas production.
Methodology: Three remote sensing instruments, namely the University of Colorado AMAX-DOAS, NOAA TOPAZ lidar, and NCAS Doppler lidar were deployed aboard the NOAA Twin Otter during summer 2010. Production rates of nitrogen dioxide (NO$_2$) and O$_x$' (background corrected O$_3$ + NO$_2$) were quantified using the horizontal flux divergence approach by flying closed loops near Bakersfield, CA. By making concurrent measurements of the trace gases as well as the wind fields, we have reduced the uncertainty due to wind field in production rates.
Results: We find that the entire region is a source for both NO$_2$ and O$_x$'. NO$_2$ production

---

Corresponding author: Email: rainer.volkamer@colorado.edu;

is highest over the city (1.35 kg hr^{-1} km^{-2} NO$_2$), and about 30 times lower at background sites (0.04 kg hr^{-1} km^{-2} NO$_2$). NO$_x$ emissions as represented in the CARB 2010 emission inventory agree well with our measurements over Bakersfield city (within 30%). However, emissions upwind of the city are significantly underestimated. The O$_x$' production is less variable, found ubiquitous, and accounts for 7.4 kg hr^{-1} km^{-2} O$_x$' at background sites. Interestingly, the maximum of 17.1 kg hr^{-1} km^{-2} O$_x$' production was observed upwind of the city. A plausible explanation for the efficient O$_x$' production upwind of Bakersfield, CA are favorable volatile organic compound (VOC) to NO$_x$ ratios for O$_x$' production, that are affected by emissions from large oil and natural gas operations in that area.

Conclusion: The NO$_2$ and O$_3$ source fluxes vary significantly, and allow us to separate and map NO$_x$ emissions and O$_x$ production rates in the Central Valley. The data is probed over spatial scales that link closely with those predicted by atmospheric models, and provide innovative means to test and improve atmospheric models that are used to manage air resources. Emissions from oil and natural gas operations are a source for O$_3$ air pollution, and deserve further study to better characterize effects on public health.

Keywords: Active and passive remote sensing; LIDAR; AMAX-DOAS; fluxes, air pollution;

1. INTRODUCTION

Ozone (O$_3$) and nitrogen oxides (NO$_x$ = NO + NO$_2$) are trace gases that are important components of air pollution. Health concerns of O$_3$ and NO$_2$ are related to respiratory illnesses such as chest pain, reduced lung function, asthma, emphysema whereas environmental concerns include reduced vegetation growth and acid rain. Due to these concerns both trace gases are recognized as air pollutants by air quality regulating agencies around the world, and are regulated by air quality standards and guidelines. The World Health Organization (WHO) air quality guideline recommends the standard to be set at 100 µg m^{-3} (~51 ppb) for O$_3$ (8 hour mean) and 40 µg m^{-3} (~21 ppb) for NO$_2$ (annual mean) [1]. The National Ambient Air Quality Standard (NAAQS) set by U.S. Environmental Protection Agency are 75 ppb for O$_3$ (8 hour maximum) and 53 ppb for NO$_2$ (annual mean) [2]. Similarly, the current air quality standard for the European Union are 120 µg m^{-3} (~61 ppb) for O$_3$ (8 hour maximum) and 40 µg m^{-3} (~21 ppb) for NO$_2$ (annual mean) [3]. Further, O$_3$ is a greenhouse gas that is relevant to climate discussions [4]. The lifecycles of O$_3$ and NO$_x$ are intimately coupled, because NO$_2$ photolysis by sunlight drives photochemical O$_3$ production, while emissions of NO destroy O$_3$ to form NO$_2$. The sum of O$_3$ and NO$_2$ is called O$_x$, and is a conserved quantity as it implicitly accounts for the destruction of O$_3$ by NO (O$_3$ titration). Excess O$_x$ is formed from the oxidation of volatile organic compounds (VOCs) in the presence of NO$_x$ [5-7]. Fig. 1 shows a schematic diagram of photochemical O$_3$ production and evolution of NO, NO$_2$, O$_3$ and O$_x$ concentrations upwind (I), within city limits (II) and downwind (III and IV) of an urban area. Different chemistry in these regions results in the characteristic spatial patterns in NO$_x$-O$_3$ distributions depicted in Fig. 1, which are: (1) Background O$_3$ present in the upwind region (I). (2) Emission of NO$_x$ in the city limits (II), which leads to (3) O$_3$ removal via reaction with NO to produce NO$_2$ (titration reaction). (4) Photochemical production of O$_3$ from VOC/NO$_x$ chemical cycles, which dominates downwind of the city center (III) and results in O$_3$ concentrations to accumulate. Further (5) the O$_3$ concentration does no longer accumulate in some distance downwind (IV), when NO$_x$ has been oxidized to NO$_y$. NO$_y$ is efficiently deposited or lost to aerosols resulting in insufficient NO$_x$ to drive VOC-NO$_x$ chemical cycles (NO$_x$ limited region). Ultimately O$_3$ removal by photolysis and dry deposition leads to a slowly decreasing O$_3$ concentration here [6-8].

Due to their importance for air quality and human health, NO_2 and O_3 plumes from point sources and urban areas have been extensively studied. Previous studies have estimated NO_2 emission rates from point sources like power plants [9], urban areas [10-13], O_3 production rates in urban plumes [14,15], the amount of O_3 transported from urban areas and its impact on regional background O_3 [15], and the relationship between O_3, NO and NO_2 as function of NO_x in urban areas [16]. However, despite decades of research, models that predict O_3 formations have not been constrained by observations at the scale of cities and immediately downwind of cities. The comparison at the local scale is important, because of uncertain and changing emission of VOCs [17], NO_x [18], complicated transport [19-21] over cities and downwind of cities, and also uncertainties in non-linear chemistry that couples VOCs, NO_x and O_3. Such chemistry is heavily parameterized in current atmospheric models used to predict O_3. The net O_3 production by VOC oxidation is related to the conversion of NO to NO_2 by organic peroxy-and hydro peroxy radicals that are formed during the airborne oxidation of VOCs by atmospheric oxidants like OH, NO_3, O_3, and Cl radicals [22]. Under high NO_x conditions, the rate of O_3 production is limited by the availability of VOCs, while availability of NO_x controls the rate of O_3 production under low NO_x conditions [22-25]. For example, the testing of detailed chemical mechanisms of VOC oxidation using simulation chamber data [26,27], and field observations [28-31] often predict lower O_3 formation rates than that are actually being observed. The uncertainty in the chemistry of O_3 formation can be of similar relevance as uncertainties in emissions, and transport [30]. Further, transport of O_x across city, state and international borders causes possible non-attainment of O_3 levels at sites downwind [15,20,32] and the changing boundary conditions complicates enforcement of regulations.

Over the course of the last decade, emission control policies aimed at reducing ambient O_3 levels have resulted into NO_x reductions in North America and Europe [18,33-38]. NO_x sources in the troposphere are primarily related to anthropogenic emissions from on-road motor vehicles and power plants. With more than half of the world population now living in urban areas, cities have developed into hotspots for NO_x sources [18,33,34] and provide opportunities for NO_x reductions that are relevant on the global scale. This trend towards urbanization on global scales is unique in the history of mankind, and has the potential to change the planet. There is an increasing need for the development of analytical approaches that are effective at quantifying emissions of NO_x, provide experimental constraints to O_x production rates, and transport in order to refine atmospheric models that are used to manage air resources.

The primary objective of this study is to demonstrate the potential and feasibility of integrated use of passive and active remote sensing instruments and column observations to estimate the rate of NO_x emissions, and investigate the O_x production from an urban area. We use the mass conservation approach to estimate source strength for NO_2, and O_x from an urban area. Recently, ground based mobile differential optical absorption spectroscopy (DOAS) measurements have been used to estimate NO_x emissions from urban areas using this approach [10,11,13]. A similar approach has also been used to probe NO_x emission from megacities using satellites [12]. We have made simultaneous measurements of NO_2 vertical columns, O_3 and wind profiles for the first time from a research aircraft. The data set provides an opportunity to estimate production of individual species and investigate the conserved quantity, O_x, which could be significantly impacted by O_3 titration in NO_x source areas such as city centers. As a case study, data from a research flight on June 15, 2010 over Bakersfield, California is presented.

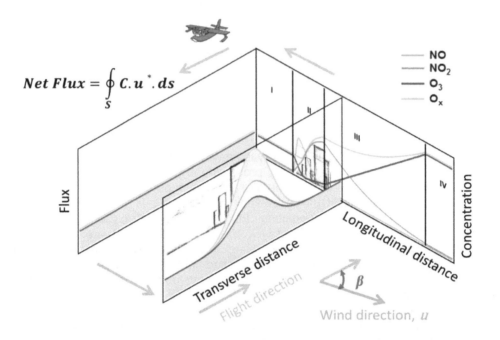

$$Net\ Flux = \oint_S C.u^*.ds$$

Fig. 1. Schematic diagram showing cross-section of ozone formation in an urban area under steady wind conditions and horizontal flux divergence measurements in a closed loop for source strength calculations. Evolution of NO (orange), NO₂ (blue), O₃ (maroon) and Oₓ (green) over different urban regions: (I) upwind, (II) urban center, (III) downwind and (IV) further downwind are also illustrated.

2. METHODOLOGY

We use a mass conservation approach to estimate the emission and production source strength of NO_2 and O_x. Neglecting the molecular diffusivity term in the mass conservation equation, the NO_2 and O_x source strengths within a given volume can be estimated from their time rate of change within the volume and the horizontal flux divergence across the boundaries enclosing the volume. We have conducted measurements of vertical columns of NO_2, O_3, and wind profiles aboard a research aircraft that flew box patterns over and near an urban area. Fig. 1 shows a conceptual schematic illustrating our approach for measuring NO_2 and O_x production rates.

Three remote sensing instruments namely (1) the University of Colorado Airborne Multi-Axis Differential Absorption Spectroscopy instrument (CU AMAX-DOAS), (2) the National Oceanic and Atmospheric Administration (NOAA) Tunable Optical Profiler for Aerosol and Ozone (TOPAZ) lidar and (3) the National Center for Atmospheric Science (NCAS) Doppler lidar were deployed aboard the NOAA Twin Otter research aircraft. The configuration of the three instruments aboard the Twin Otter is shown in Fig. 2. A total of 52 research flights were conducted over the course of two months (May 19-July 19, 2010) as part of the California Research at the Nexus of Air Quality and Climate Change (CalNex) [39] and the Carbonaceous Aerosol and Radiative Effects Study (CARES) [40] field campaigns in California during summer 2010. Most of the flights were focused on the Los Angeles basin and Greater Sacramento area. More details on the individual Twin Otter research flights can be found in Ryerson et al. [39]. One of the foci of this deployment was to constrain the emission and production of NO_2 and O_3 upwind, within and downwind of urban areas.

2.1 AMAX-DOAS

The CU AMAX-DOAS instrument [41,42] uses scattered sunlight as the light source (passive remote sensing). The scattered sunlight spectra are analyzed for the presence of absorbers like NO_2, glyoxal (CHOCHO), formaldehyde (HCHO) and oxygen dimer (O_4) among others using the DOAS method [43]. The instrument and its performance during CalNex and CARES field campaigns are described in detail in Baidar et al. [42]. Briefly, a telescope pylon is mounted on the outside of the window plate of the aircraft and includes a rotatable prism to collect scattered photons from different elevation angle (EA) i.e. angle relative to the horizon. Spectra collected from different EA contain information from different layers in the atmosphere and hence can be used to obtain information about vertical distribution of trace gases. The collected photons are transferred to a spectrometer / charge coupled device (CCD) detector system via optical glass fiber bundle. Here we will only present data from nadir viewing geometry from the flight over Bakersfield.

The measured spectra were analyzed, for NO_2 in a wavelength range from 433 to 460 nm, against a fixed zenith reference spectrum recorded during the same flight in a clean environment and flying at relatively high altitudes (3 – 5 km; 3.5 km for this flight). Zenith spectra were recorded frequently, and are used to correct for stratospheric NO_2 contributions and NO_2 above the aircraft. The nadir NO_2 differential slant column densities (dSCDs) are observed below the plane and correspond to the average integrated difference in concentration of the absorber along the light path with respect to the reference. Since most of the NO_2 sources in an urban environment are located close to the surface, the retrieved nadir dSCD was considered to be the boundary layer slant column (dSCD$_{bl}$). NO_2 nadir measurements were performed every 20-25 s and hence NO_2 data points are available every ~1.5 km horizontally. For the conversion of nadir NO_2 dSCD$_{bl}$ into boundary layer vertical column densities, VCD$_{bl}$, the geometric Air Mass Factor (AMF$_{geo}$) approximation was applied.

$$AMF_{geo} = \frac{1}{1+\frac{1}{\cos(SZA)}} = \frac{SCD_{bl}}{VCD_{bl}} \qquad (1)$$

Here, SZA refers to solar zenith angle at the time of the measurement. This approach is in good agreement with explicit radiative transfer calculations for California while flying between 2 and 4 km. Radiative transfer calculations for the conditions of the Bakersfield case study (flight altitude: 2 km, SZA: <25°), and comparisons with ground based vertical columns consistently reveal the uncertainty in AMF$_{geo}$ to be less than 7% [42]. The overall uncertainty in NO_2 VCD for the Bakersfield case study is estimated to be around 9% (AMF$_{geo}$: <7%, NO_2 cross-section: ~5%, DOAS fit: ~3%) [42,44].

2.2 TOPAZ

NOAA's nadir-looking TOPAZ differential absorption lidar is a compact, solid-state-laser-based O_3 lidar that emits pulsed laser beams at three tunable wavelengths in the UV spectral region between about 285 and 300 nm [45]. The differential attenuation of the three wavelengths due to O_3 permits the retrieval of O_3 concentration profiles along the laser beam path [46]. TOPAZ O_3 profiles were computed every 10 s (or about 600 m horizontally) with a vertical resolution of 90 m. The ozone profiles extend from about 400 m beneath the plane to near the ground. O_3 values in the lowest two measurement bins (lowest 180m) above ground level (AGL) are typically not used because of poor signal to noise ratio.

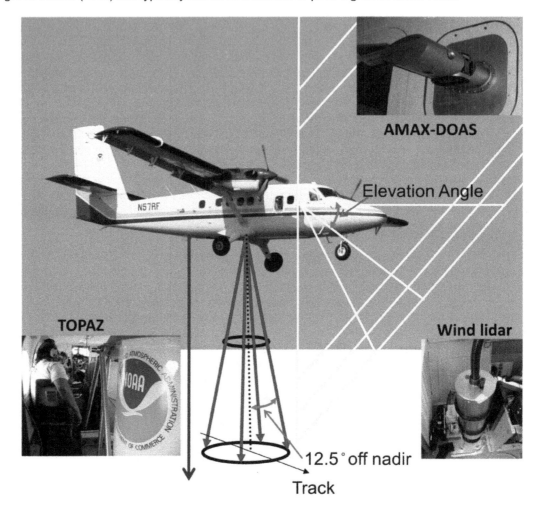

Fig. 2. Instrumental setup of CU AMAX-DOAS, NOAA TOPAZ lidar and NCAS Doppler wind lidar aboard the NOAA Twin Otter research aircraft during CalNex and CARES field campaigns. The yellow, purple and maroon lines represent viewing geometry of CU AMAX-DOAS, NOAA TOPAZ lidar and NCAS Doppler wind lidar respectively. The three instruments are also shown in the insets.

The TOPAZ lidar also provided aerosol backscatter profiles for the longest (and least absorbed by O_3) of the three emitted wavelengths near 300 nm. The time resolution of the aerosol backscatter profile measurements is the same as for O_3, but the vertical resolution is much finer at 6 m. We used these highly resolved lidar backscatter profile data to retrieve boundary layer height (BLH) by employing a Haar wavelet technique [47]. This approach is based on the (often valid) assumption that the aerosol concentration is higher in the boundary layer (BL) than in the lower free troposphere (FT). The altitude at which the strongest aerosol gradient is found by the wavelet technique is used as an estimate of the

BLH. At times, the contrast in aerosol backscatter between the BL and the overlying FT is not sufficient to yield reliable results, and the BLH is not reported for such scenarios.

We used the BLH estimates and O_3 profiles measured with the TOPAZ lidar to compute O_3 column data integrated over the depth of the BL. To fill in data gaps in the O_3 profiles close to the ground, we averaged the ozone measurements in the lowest two gates with usable data (typically 200 - 300 m AGL) and extrapolated this value to the ground. We then integrated these extrapolated ozone profiles from the surface to the top of the BL to yield BL O_3 column density along the flight track at 10-s resolution. When BLH estimates were not available from the backscatter profile for a given O_3 profile, BLH was interpolated from adjacent measurements to compute O_3 vertical column over the BLH. TOPAZ O_3 measurements have been extensively compared to and agree well (±2-9%) with in situ airborne O_3 observations [48].

2.3 Doppler Wind Lidar

Information on the wind structure below the aircraft was provided by the NCAS Doppler lidar [49] mounted in the Twin Otter cabin. The lidar measures the Doppler shift of radiation scattered from atmospheric aerosol particles to estimate the component of wind along the lidar line of sight. Typical precision of the lidar radial wind measurements under acceptable aerosol loading is better than a few tens of cm s^{-1}. The lidar was mounted in the cabin with the beam transmitted vertically through a small camera port located on the underside of the aircraft (Fig. 2). In order to measure the horizontal component of the winds a rotatable refractive wedge mounted in the port directed the beam to 12.5° off nadir. The original scanner design included two wedges, which provides greater beam deflection and enables vertical pointing; however poor optical quality of the wedges forced us to eliminate the second wedge to reduce total attenuation through the scanner.

During flight operations the wedge was rotated to four different azimuth angles (45, 135, 225, 315°) relative to the flight track. Dwell time at each azimuth angle was 1 s for most of the Doppler measurements during CalNex. A complete rotation among the four azimuths required 8 s, including the time required to rotate the wedge to a new position. At the nominal Twin Otter speed of 60 m s^{-1} a complete 4-beam scan was completed about every 500 m. Vertical resolution of the lidar wind measurements was roughly 50 m.

Information on aircraft speed and orientation was obtained from the CU AMAX-DOAS motion compensation system [42]. Additionally, we used the surface return at the four look angles, for which the only Doppler shift results from motion of the aircraft, to provide additional information on aircraft orientation and velocity. For the case described here, a 19-beam running average of the radial wind estimates at each of the four azimuth angles was computed to improve precision of measurements. After removal of the Doppler shift induced by aircraft motion, the velocities from the four azimuth angles were combined in a least-squares type algorithm to estimate the mean wind speed and direction in each of the 50 m range gates where backscatter was high enough to provide a strong signal. The wind speed and direction were averaged up to the BLH before further calculations of horizontal flux. The uncertainty in the wind measurement is estimated to be around 6% based on the difference between wind retrievals from a longer (19-beam) and a shorter (3-beam) running average wind fits.

2.4 Bakersfield Case Study

Bakersfield is a city located in the southern part of the Central Valley, CA, surrounded mostly by agricultural area and oil and natural gas operations. In the summer months, wind blows predominantly from the northwest down the valley providing steady wind conditions necessary for the method presented here. The Bakersfield area also makes for an interesting case study to probe NO_2 and O_x production from a large city influenced by intense agriculture and petrochemical production. In particular, we have probed (i) background air unaffected by urban anthropogenic emissions, (ii) air upwind, influenced by agricultural and petrochemical production, (iii) urban emissions from the city, and, (iv) the chemical evolution downwind, after it is perturbed by urban emission inputs.

The flight plan of the Twin Otter on June 15, 2010 (see Fig. 3) was designed to interrogate NO_x emissions and constrain the O_3 production from different source regions enclosed by "boxes" by applying the horizontal flux divergence approach. The flight plan included an enclosed box, over areas with no major emission sources, in the northwest of Bakersfield to characterize the background conditions (Box A; see Fig. 4). Two boxes (Box C and D; D is twice the size of C) were flown over the city of Bakersfield to constrain emissions/productions from the city. In order to contrast NO_2 and O_3 production upwind and downwind of Bakersfield, two additional boxes (Box B and E) were created by interpolating the measured NO_2, O_3 and wind data for the western legs (shown as diamonds in Fig. 4). A larger trapezoid (Box F) was flown, connecting the three boxes, and enclosing the entire greater Bakersfield region. It took approximately 15, 13 and 18 minutes to complete boxes A, C and D respectively while the larger Box F took ~75 minutes. The entire box patterns were flown at a constant altitude (~2000 m above sea level), well above the BL. Details related to the times and meteorological conditions encountered for each boxes are summarized in Table 1.

Table 1. Meteorological conditions on June 15, 2010 for the closed boxes flown near Bakersfield, CA

Box	Area (km x km)	Time (UTC)	Wind speed (m/s) mean ± sdev	Wind direction (°) mean ± sdev	Boundary layer height (m, AGL) min	max
A	20 x 14	20:24-20:38	5.6 ± 1.7	283 ± 44	1005	1548
B	20 x 19		4.3 ± 1.6	314 ± 41	760	1425
C	20 x 9	19:41-19:54	4.1 ± 1.5	317 ± 49	936	1425
D	20 x 18	21:07-21:25	4.7 ± 0.9	317 ± 26	1049	1481
E	20 x 18		3.0 ± 0.8	328 ± 27	1066	1450
F	67 x 56	19:54-21:11	3.3 ± 1.8	313 ± 41	760	1524

2.5 Horizontal Flux and Source Strength

For each transect, the gas flux at a point, x along the flight path is obtained by multiplying the column measurement at that location, *column(x)* by the corresponding wind speed averaged over the BLH, $u^*_{avg}(x)$ [10,11]. The flux calculation through a surface area, A is shown in equation 2:

$$\int \vec{J}.d\vec{A}_{area} = \int_{x1}^{x2}\int_{0}^{BLH} conc(z).u^*(z).dzdx = \int_{x1}^{x2} u^*_{avg}(x).column(x)dx \quad (2)$$

Where J corresponds to flux at any location, x to the flight direction, z to the altitude, u^* to the wind speed orthogonal to the flight direction (x) and is assumed to be constant over the BLH, $\dot{u}^*_{avg}$ to wind speed averaged up to the BLH, and

$$column = \int_0^{BLH} conc(z).dz \qquad (3)$$
$$u^* = u.\sin(\beta) \qquad (4)$$

Here u is wind speed and β is angle between wind direction and flight heading.

Flux measurement in a closed loop can be used to estimate source strength within the enclosed volume [10,11]. The general continuity equation in the integral form is given in equation (5) and is the basis for the source strength calculation. It involves three terms: source, flux divergence and rate of change of concentration.

$$Q_{net} = \oint_S \vec{J}.d\vec{A}_{area} + \int_{Vol} \frac{\partial conc}{\partial t} dV_{vol} \qquad (5)$$

i.e. the net source strength of an enclosed volume, Q_{net} is the sum of fluxes through all areas along the closed loop (incoming and outgoing) and change in concentration inside the volume.

We assumed that the time dependence of concentration in the enclosed volume is zero over the time scale of our measurement. Our measurements were performed during the midday when rate of change of NO_2 and O_3 concentration in the Bakersfield area is very small (see Fig. 5). Hence, we neglected the second term on the right hand side in equation (5). We also assumed that the net vertical exchange and deposition are negligible over the timescale of our measurement. Hence, the net flux i.e. the difference in fluxes entering and leaving the enclosed volume through the walls gives the source strength for the species of the particular enclosed area at the time of the measurement.

2.6 Daily NO$_x$ Emission

Daily NO_x emission was estimated based on the computed NO_2 source strength, diurnal profile of NO_2 and the NO_x to NO_2 ratio measured at the California Air Resource Board (CARB) monitoring station at Bakersfield. It is computed as:

$$NO_x \, emssion = \sum_{t=10}^{20} E.\frac{[NO_2(t)]}{[NO_2(t_0)]}.\frac{[NO_x(t)]}{[NO_2(t)]} \qquad (6)$$

Where E is the computed NO_2 production rate from (5), t is hour of the day, t_0 is the hour of our measurement. Therefore, the daily NO_x emission is the sum of product of measured NO_2 emission rate at time t_0, ratio of NO_2 at time t and t_0 and ratio of NO_x to NO_2 at time t over the course of the day. In order to minimize potential bias created due NO_x accumulation overnight, we only calculated daily NO_x emission for the period when NO_2 and NO_x measurements at the CARB station were stable (10:00-20:59 PST). Fig. 5 shows NO_2, NO_x and O_3 mixing ratios measured hourly at the Bakersfield CARB monitoring station on that day. The time period of our measurement and time frame for the daily NO_x emission calculation are also shown in Fig. 5.

3. RESULTS AND DISCUSSION

The measurement on June 15, 2010 over Bakersfield, CA was performed at mid-day when the change in NO_2 and O_3 concentration is very small, providing chemically stable conditions most suitable for source strength calculations. This is supported by NO_2 and O_3 measurements at the CARB monitoring station at Bakersfield (see Fig. 5). Column O_3, BLH from the NOAA TOPAZ lidar, column NO_2 from the CU AMAX-DOAS, and wind speed and direction from NCAS Doppler lidar are shown in Fig. 3. Fig. 3a shows the BLH at the time of the measurement retrieved from TOPAZ backscatter profiles. In general, the highest O_3 and NO_2 VCDs were measured in the southeastern corners of the boxes especially for boxes with significant emission sources (Fig. 3b and 3c), consistent with the prevailing wind conditions. We calculated NO_2 and O_x production rates for the six boxes.

Fig. 3. Maps of (A) Boundary layer height (BLH) above ground level, (B) O_3 vertical columns up to the BLH, (C) NO_2 vertical columns and (D) wind speed and direction from the flight over Bakersfield, CA on June 15, 2010. BLH and O_3 columns were measured by NOAA TOPAZ lidar, NO_2 vertical columns by CU AMAX-DOAS and wind speed and direction by NCAS Doppler wind lidar. Black diamond on A shows the location of CalNex Bakersfield supersite.

The wind speed and direction and BLH during the time of measurements for different boxes are given in Table 1. The wind was blowing predominantly from the northwest and provided ideal conditions as horizontal flux divergence measurements require steady wind fields (also see Fig. 3d). The variability in wind speed and direction was larger for upwind boxes (A and B) compared to downwind boxes (D and E). The BLH range for boxes D and E were also smaller compared to other boxes. The observed BLH variability is most likely a combination of land use changes (irrigated fields vs. dry land vs. urban heat island) and the fact that upslope flow over the foothills east of Bakersfield favor BLH growth, while strong subsidence over the middle of the valley acts to suppress BL growth. We use the BLH variability to estimate the amount of BL air column susceptible to exchange with the FT (see Section 3.1.). Since all the parameters needed to quantify flux are constrained by measurements here, the horizontal variability in BLH does not pose a limitation to our approach. Notably, the variability of BLH remains difficult to predict by atmospheric models, and warrants further investigation.

Background corrected O_x (O_x' from here on) column up to the BLH, computed as the sum of NO_2 and background corrected O_3 columns, is shown in Fig. 4. Background correction for O_3 is needed as we are interested in the source strength of the area at the time of measurement i.e. the amount that is being produced locally. The background correction also minimizes any potential biases due to BL-FT exchange in case of strongly varying BLH. If BLH is constant over a box, then background correction is not necessary, because incoming and outgoing background fluxes are the same and cancel each other. Note that we measure column amount of O_3 and NO_2. Background O_3 levels were calculated based on the mean O_3 concentration over the BLH in the northwestern corner of the Box A (see Fig. 4). We note that 'background' air in the Central Valley is affected by transport of pollution emitted upwind; indeed we find significant production of O_x' in Box A. However, low and similar amounts of NO_2 are transported into and out of Box A, and net production of measured species is the lowest observed anywhere. We find no evidence for major emission sources within Box A. The mean and the standard deviation of the background O_3 concentration was $1.20 \pm 0.03 \times 10^{12}$ molecules cm^{-3} (48.8 ± 1.2 ppbv, 1 ppbv $= 2.46 \times 10^{10}$ molecules cm^{-3}). It was assumed that this background O_3 concentration is representative of the entire area. A background O_3 column was calculated for each measurement point along the flight track by multiplying this O_3 concentration with BLH at that location. This background column was then subtracted from the measured O_3 column to determine O_3', which corresponds to the excess O_3 column at each point along the flight track. NO_2 VCDs were used without further corrections as they were below the detection level over that area (4.2×10^{14} molecules cm^{-2}, ~130 pptv). The average column NO_2 to O_x' partition ratio increased from 2% over the background site to 7% over the city of Bakersfield. Thus, NO_2 gas forms a significant portion of O_x' over the urban area and would result in a bias if O_3 production rates were calculated instead of O_x. By investigating O_x' we eliminate the uncertainty due to titration of O_3 by NO to form NO_2 in the NO_x source regions.

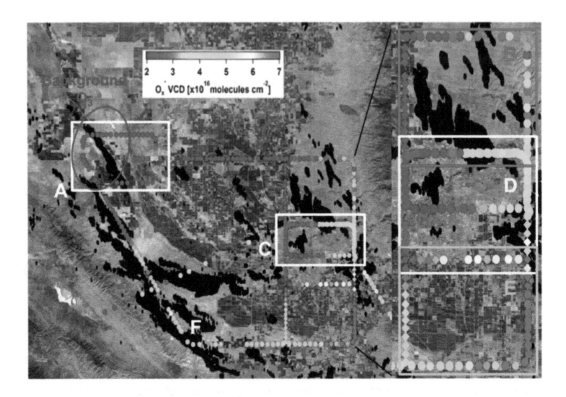

Fig. 4. Map of O_x' vertical columns up to the BLH. Rectangles represent different boxes flown over Bakersfield: (A) upwind background area, (C) over the city, and (F) over the entire area. Colored rectangles in the inset highlight three boxes: (B) upwind, (D) over the city, and (E) downwind. Box B and E were created for comparison purposes by interpolating the western legs. Interpolated data are shown as diamonds. Black areas in the map show active oil and gas wells in the region.

The minimum, maximum and mean mixing ratios of NO_2 and O_3 for each box are also given in Table 2. The average NO_2 and O_3 concentrations were lowest for the background box. The NO_2 concentrations show higher variation within a box as well as between the boxes, indicating highly local NO_2 sources in the area. The mean O_3 does not vary much between the boxes (see Table 2). Notably, the maximum O_3 concentrations were observed to be generally related to the O_x' production rates in the box upwind of a given box, which is expected. Both O_3 and NO_2 showed the highest average concentration over the downwind box (E). Considering that the production rates are lower compared to the boxes upwind (B and D), there could be some accumulation of NO_2 and O_3 taking place in this box.

The enclosed areas are sources for both NO_2 and O_x' for all the boxes investigated. The NO_2 and O_x' production rates calculated for different boxes are given in Table 2. The production rates were calculated using equation (5) and have been normalized for the area of the boxes so that they can be directly compared to each other. As expected, the background, Box A, has the lowest production rate for both NO_2 and O_x'. The NO_2 production rate in the background box was 0.04 kg hr^{-1} km^{-2}. The NO_2 production rate was highest for Box D and amounts to 1.35 kg hr^{-1} km^{-2} above Bakersfield. This is consistent with the present knowledge that urban city limits are the dominant source for NO_x emissions in California [18].

Boxes C and E, located in the northern half of Box D, and immediately downwind of Box D respectively, show about 13 times lower NO_2 production rate, indicating that the NO_2 sources from the urban area are highly localized. Interestingly, the site upwind of Bakersfield (Box B) was also found to be a significant source for NO_2 compared to the downwind site (Box E).

Table 2. NO_2 and O_3 mixing ratios and NO_2 and O_x' production rates normalized by area of the boxes for each box near Bakersfield, CA on June 15, 2010

Box	Mixing ratio[1]		Production rates[4]	
	NO_2 (pptv)	O_3 (ppbv)	NO_2[2]	O_x'[3]
	min / max / mean	min / max / mean	$(\times 10^{-2}\ kg\ hr^{-1}\ km^{-2})$	$(kg\ hr^{-1}\ km^{-2})$
A	22 / 864 / 298	45 / 59 / 51	4 ± 8	7.4 ± 0.6
B	38 / 951 / 497	47 / 62 / 58	60 ± 6	17.1 ± 0.8
C	145 / 2425 / 852	47 / 67 / 58	11 ± 15	11 ± 2
D	149 / 1554 / 694	52 / 68 / 60	135 ± 12	13 ± 1
E	563 / 1948 / 1183	56 / 76 / 66	12 ± 10	11 ± 1
F	22 / 1948 / 582	45 / 76 / 59	39 ± 1	9.4 ± 0.1

[1]Mixing ratio is calculated assuming that the NO_2 and O_3 are uniformly distributed over the boundary layer. Conversion: 1 pptv = 2.46 x 10^7 molecules cm^{-3} and 1 ppbv = 2.46 x 10^{10} molecules cm^{-3}.
[2]Molecular weight of NO_2 (MW_{NO2}) =46 g/mole
[3]Molecular weight of O_x ($MW_{Ox'}$) = 48 g/mole
[4]Error in the production rates represents total propagated measurement uncertainty. Details are provided in Section 3.1.

The O_x' production rate for the background box was 7.4 kg hr^{-1} km^{-2}, the lowest of all boxes. Box A likely represents the O_x' production rates for regions in the Bakersfield area that are not affected directly by the urban emissions. Notably, the NO_2 production from within box A was the lowest we have observed in this case study. However, our approach does not attempt to make a correction for NO_2 losses due to photochemistry and deposition, and as such the reported NO_2 production has to be considered a lower limit. While the measured NO_2 flux was essentially zero within error of the measurements, this indicates that comparable amounts of NO_2 enter and exit the box, and that enough NO_x was present to produce O_3. This was confirmed by our observations of elevated O_x production in box A. Interestingly, the O_x' production rates over the Bakersfield city limit (Box C and D) and downwind site (Box E) only showed small enhancement (< factor of 1.75) over the background O_x' production rate. This indicates that even though the NO_2 levels in the wider area surrounding Bakersfield are relatively small (~330 pptv), there is enough NO_x to sustain photochemical O_x production in the entire region. Surprisingly, the upwind box (Box B) was found to have the largest O_x' production rate (17.1 kg hr^{-1} km^{-2}, 2.3 times that of Box A). The O_x' production rate in Box B was 133% that of urban Box D while the NO_2 production was only about 40% of the urban box. A plausible explanation for our observation of efficient and high O_x' production from Box B could be from enhanced VOC levels due to large oil and natural gas operations in the area, creating favorable conditions for enhanced O_3 production (high VOC/NO_x ratio). Oil and natural gas production is a source for atmospheric methane, a greenhouse gas, and other more reactive hydrocarbons as well as NO_x. The observed elevated emissions of NO_x in box B indicate emissions are active in this area. While higher NO_2 is likely to contribute to the higher O_3 production rate, additional VOC emissions are needed to explain such a high increase in the O_3 production rate. We are unable to conclude about additional VOC sources from our data at this point, but note that some emissions of reactive hydrocarbons are expected from the oil and natural gas production in the area that could help accelerate O_3 formation. Notably, the O_3 produced within Box B was only partially transported into the downwind boxes, where lower O_3 production rates were observed. This decrease in O_3 production rates downwind of Box B are probably related to higher NO_2 concentrations, and a different VOC/NO_x ratio as air mixes with urban sources. The net effect of the added emissions from urban sources was a lowering of the O_3 production rates. The black areas in Fig. 4 represent active oil and gas wells in the region [50].

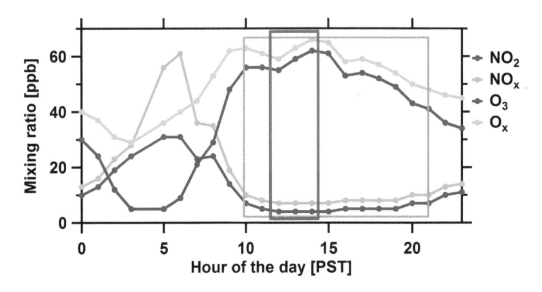

Fig. 5. Diurnal variation of NO_2, NO_x, O_3, and O_x measured at the Bakersfield CARB monitoring station on June 15, 2010. The red and grey rectangles represent the timeframe of our measurement and time period used for daily NO_x emission calculation respectively.

The measured NO_2 production rates were used along with the diurnal profile of NO_2 and NO_x measured at the CARB monitoring network station of Bakersfield (see Fig. 5) to provide an estimate of daily NO_x emissions in Table 3. We only considered the daylight hours (10:00-20:59 PST) when the measured NO_2 and NO_x at the CARB station were stable in order to minimize potential bias due to NO_x accumulation overnight. The uncertainty in Table 3 only considers error in measured production rates and does not include spatial and temporal variability in NO_x and NO_2 across the region. The daily NO_x emission from Bakersfield was estimated to be around 10.7 metric tons for June 15, 2010 from 10:00-20:59 PST, compared with 13.5 tons NO_x for the same time frame in CARB 2010 emission inventory (CalNex-2010 modeling inventory) [51]. There is a mismatch in the location of NO_x emissions within Bakersfield. Our measurements suggest that the large portion of the emission occurs in the southern half of Box D. Note that NO_x emissions of Box C were also part of Box D (i.e., form the northern half of Box D). Emissions in Box C were comparatively very small. In contrast, emissions from Box C form a significant portion of overall emission of Box D in the CARB 2010 emission inventory. The NO_x emissions for the entire study area (Box F: 32.1 tons) were comparable to those in the CARB 2010 emission inventory in the same area (29.2 tons). However, there are differences in the locations of the NO_x emissions here as well. The background NO_x emissions are higher in the emission inventory whereas emissions over the oil and natural gas operations are significantly underestimated. Table 3 also compares hourly emission rates for the hour of our measurement. The measured emissions were lower for all the boxes except for the upwind box, B. Considering that the daily emission estimates are in better agreement compared to the hour of measurement, there is a discrepancy in the timing of NO_x emission in the emission inventory. The diurnal profile of NO_2 and the NO_x to NO_2 ratio varies between days as well as seasons and hence we do not attempt to scale up to the yearly NO_x emission. However with regular flights over different times of the day and course of different seasons, the combination of active and passive remote sensing has the potential to constrain and improve NO_x emissions in emission inventories.

Table 3. Daily and hourly NO_x emissions calculated for June 15, 2010 from (i) using derived NO_2 production rate and NO_x and NO_2 diurnal profiles measured at CARB monitoring station at Bakersfield, CA and (ii) NO_x emissions from CARB 2010 emission inventory for that day. Errors represent error due to uncertainty in NO_2 production rates

Box	NO_x Emissions[1] (metric tons)			
	This work		CARB 2010 emission inventory	
	Daily[2]	Hourly[3]	Daily[2]	Hourly[3]
A	0.2 ± 0.5	0.02 ± 0.04	1.1	0.11
B	5.1 ± 0.5	0.40 ± 0.04	1.1	0.11
C	0.4 ± 0.6	0.03 ± 0.05	10.6	1.05
D	10.7 ± 0.9	0.85 ± 0.07	13.5	1.33
E	0.9 ± 0.8	0.07 ± 0.06	2.2	0.22
F	32.1 ± 0.9	2.56 ± 0.07	29.2	2.91

[1]*Molecular weight of NO_x (MW_{NOx}) = 46 g/mole.*
[2]*Daily = 10:00-20:59 PST.*
[3]*Hourly = hour of our measurement*

3.1 Error Estimates

Error in calculated fluxes and source strengths are a function of uncertainties in the measurements of individual species, winds, uncertainties about sinks (dry deposition and oxidation), and the variability of atmospheric state. Previous source strength calculations have found the uncertainty in the wind to be the largest source of error as it was not measured concurrently [10,11]. The uncertainty in the wind measurements is estimated to be around 6%. Thus, the uncertainty in the wind measurement itself has a relatively small effect on the production rates in our study. This uncertainty is very likely not representative of wind variability within the boxes but the variability in wind are captured as part of individual wind measurements.

The overall uncertainty in NO_2 VCD is ~9%. The contributions of different error sources in NO_2 VCD uncertainty is given in Section 2.1. It is assumed that all of NO_2 were located inside the BL. Based on the amount of NO_2 above the BL in the vertical profile through the city center (see Fig. 6) we estimate this leads to a systematic error of around a few percent. The lidar O_3 profile measurements at 90 m vertical resolution have an error of typically 6-10% and can be improved by integration and averaging [45,48]. O_3 data were integrated vertically (~10 points) and horizontally (2-3 points) for flux calculations at each location. As a result, the statistical uncertainties in the integrated O_3 are reduced to ~2%. O_3 data to the ground were extrapolated from measurement at lowest two gates assuming a well-mixed BL (see section 2.3). Based on the difference between measurement at the CARB stations and extrapolated values, we estimate this systematic error to be around 5%. The uncertainty in BLH retrieval for TOPAZ backscatter profile is ~7%. Considering the standard deviation of background O_3 concentration (~4%), BLH uncertainty, O_3 column uncertainty and NO_2 column uncertainty, the overall error in the O_x' column is ~8%. Thus the total measurement uncertainty in the individual flux measurement is ~10-11%. The error for production rates of NO_2 and O_x' tabulated in Table 2 represents the overall uncertainty due to propagated error in individual column of the species and wind speed and direction for each box.

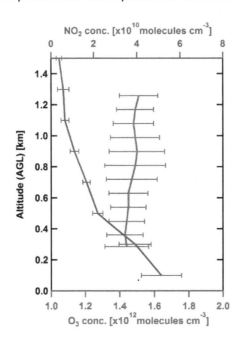

Fig. 6. Mean O_3 profile measured by NOAA TOPAZ lidar for Box D and NO_2 profile measured by CU AMAX-DOAS from a low approach at Bakersfield airport. Error bars for O_3 profile is the standard deviation of the mean and NO_2 shows measurement uncertainty.

Vertical transport, atmospheric sinks and chemical transformations over the transport time between source and sampling regions are other potential sources of errors in the measured production rates for NO_2 and O_x'. Our approach assumes that all transport through the enclosed box occurs horizontally. Entrainment flux is a product of entrainment velocity (w_e) and the difference in trace gas concentrations between the FT and the BL (ΔC) ($E_{flux} = w_e \times \Delta C$). To constrain the magnitude of potential vertical flux, we have used the rate of BLH growth as the entrainment velocity. This neglects the potential contribution of large-scale, mean vertical velocity and BLH advection. The northern legs for Box C and D overlap in location, but were flown ~90 minutes apart and provide an opportunity to calculate the BLH growth rate. It was estimated to be 1.2 cm s^{-1}. NO_2 and O_3 concentrations were determined from the vertical profiles obtained from the low approach over the Bakersfield airport (see Fig. 6). We estimate the vertical flux to be 2.30×10^{-3} kg km^{-2} hr^{-1} for NO_2 and 0.21 kg km^{-2} hr^{-1} for O_x' for Box D. Hence, the potential error due to vertical transport is likely to be smaller than 2%.

Typical dry deposition velocities, w_d, for O_3 and NO_2 in the continental environment are 0.4 and 0.1 cm s^{-1} respectively [6,7,52]. We calculated the depositional flux as $D_{flux} = w_d \times C$, where C is the trace gas concentrations at the lowest layer. This could result in error of ~

10% in the O_x' and ~1% in the NO_2 production rates. For an air mass transport time of 1 hour (between production and measurement), Ibrahim et al. [10] estimated the error in the production rate due to chemical transformation of NO_x, based on average atmospheric NO_x lifetime, to be around 10%. Considering that the transport time for our conditions is around 0.5 hour, we estimate the error due to chemical transformation of NO_2 to be less than 5%. The atmospheric lifetime of O_3 is more than 3 times that of NO_x. Hence, we estimate the error in O_x' production rate due to chemical transformation to be smaller than 2%. Thus, the potential error due to entrainment, dry deposition and chemical transformation is in the same order as the total measurement uncertainty in the production rates.

We observed very high local variation in BLH and this could also potentially result in error in the calculated production rates. The BLH variability makes the air column susceptible to exchange with the FT via horizontal transport. We have tried to bind the magnitude of this exchange in two ways: Method A calculates it as a product of the relative amount of air column (with respect to the average) susceptible to this exchange based on the difference in the average BLH measured for the upwind and downwind legs and our measured production rates; Method B calculates the same number based on the maximum and minimum BLH, assuming they occur equally frequent within each box. This is likely an upper limit estimate of such transport, since BLH is something in between most of the time. Note that we only know the BLH along the edge of the boxes and not within the boxes. We find that the uncertainty due to this potential FT exchange accounts for 1-8% (method A) of the overall horizontal flux. Method B yields 15-30% as an upper limit for FT exchange. To our knowledge the horizontally variable BLH as a mechanism for BL-FT exchange has not previously been studied, and deserves further investigation. We consider the error in horizontal fluxes from method A to be most likely representative of uncertainty in the production rates listed in Table 2 due to such an exchange.

4. CONCLUSION AND OUTLOOK

We demonstrated the feasibility of co-deployed active and passive remote sensing instruments aboard a research aircraft to study NO_x emissions and O_x production rates out of an urban area. NO_2 vertical column, O_3 vertical profile and wind profile measurements aboard the aircraft were used to calculate NO_2 and O_x' fluxes from source areas along the flight track.

The advantages to co-deployment of these three remote sensing instruments on a mobile platform for this kind of study are as follows:

1. The flux calculations are fully experimentally constrained. In particular, measurements of wind and BLH along the flight track decouple horizontal and vertical transport; column measurements integrate pollutant concentrations over the BLH, and are inherently insensitive to vertical inhomogeneity.
2. Measurements constrain NO_2, O_3, O_x (O_3+NO_2), enabling studies of NO_x emissions and O_x production rates also over NO_x source areas, i.e., under conditions when O_3 concentrations are reduced due to NO emissions (O_3 titration to form NO_2).
3. Colocation of all three sensors on a single platform minimizes error, and makes the flux calculation straightforward by assuring sampling on similar temporal and spatial scales.

The horizontal flux divergence approach presented here for a case study in the Bakersfield area has comparatively small error for the largest box (Box F) and larger error for the smallest box (Box C). The overall measurement uncertainty in the individual flux measurement is in the order of 10-11%. The potential error due to entrainment, dry deposition and chemical transformation is of a similar magnitude, and can in the future be further reduced through coupling with atmospheric models. Our measured NO_2 and O_x' production rates reveal higher O_3 production upwind of Bakersfield in an area with active oil and natural gas production. This finding is highly significant within experimental error, and spatially well separated from urban source areas. Comparison of NO_x emissions with the CARB 2010 emission inventory suggest that the NO_x emissions from the urban area are well represented in the inventory. However, the location and timing of the NO_x emissions within the urban area could be improved. In contrast, NO_x emissions over areas with active oil and natural gas production were found to be significantly underestimated; higher background emissions compensate for these local effects over the entire study area. The atmospheric impacts of emissions from oil and natural gas production deserve further investigation.

Models that predict O_3 formation have not previously been constrained by observations at the scale of cities and immediately downwind of cities. The synergistic benefit of combining active and passive remote sensing instruments demonstrated here holds great potential as an innovative tool to improve NO_x emission inventories (emitted amounts and location) as well as constrain O_x production rates experimentally, and over extended areas. The local variations in BLH deserve further investigation as to their role in the exchange of air between the BL and the FT. Further, other trace gases (e.g., formaldehyde and glyoxal) can be measured by AMAX-DOAS and hold largely unexplored potential to extend this approach to the study of VOC oxidation rates. The co-deployment of AMAX-DOAS, TOPAZ lidar and Doppler wind lidar during 51 remaining flights provide a valuable dataset to locate and constrain NO_x emissions over much of California especially the South Coast Air Basin, the Bay area, as well to assess the transport of NO_2 and O_x across the US-Mexican border.

ACKNOWLEDGEMENTS

The authors thank Stuart McKeen (NOAA ESRL, Boulder) for providing the CARB 2010 emission inventory data and the NOAA Twin Otter pilots and flight crew for their support during the campaigns. F.D. acknowledges Halo-Photonics for their help in re-configuring the NCAS Doppler lidar. The scanner for the Doppler lidar was developed with support from CIRES IRP grant to R.M.H. S.B. is recipient of a ESRL/CIRES graduate fellowship. R.V. acknowledges financial support from the California Air Resources Board contract 09-317.

COMPETING INTERESTS

The authors declare no competing interests.

REFERENCES

1.   World Health Organisation: Air Quality and Health. 2011; Available at: http://www.who.int/mediacentre/factsheets/fs313/en/. Accessed 09/25, 2013.
2.   U.S. Environmental Protection Agency: National Ambient Air Quality Standard (NAAQS). 2012; Available at: http://www.epa.gov/air/criteria.html. Accessed 09/25, 2013.
3.   European Commission: Air Quality Standards. 2013; Available at: http://ec.europa.eu/environment/air/quality/standards.htm. Accessed 09/25, 2013.

4. IPCC, 2007: Summary for Policymakers. In: Climate Change 2007: The Physical Science Basis. Contribution of Working Group I to the Fourth Assessment Report of the Intergovernmental Panel on Climate Change [Solomon, S., Qin, D., Manning, M., Chen, Z., Marquis, M., Averyt, K.B., Tignor, M., and Miller, H.L.(eds.)]. Cambridge, United Kingdom and New York, NY, USA: Cambridge University Press.

5. Haagensmit AJ. Chemistry and Physiology of Los-Angeles Smog. Ind. Eng. Chem. 1952;44(6):1342-6.

6. Finlayson-Pitts BJ, Pitts Jr JN. Chemistry of the Upper and Lower Atmosphere. San Diego, CA: Academic Press; 2000.

7. Seinfeld JH, Pandis SN. Atmospheric Chemistry and Physics: from air pollution to climate change. 2nd ed. Hoboken, New Jersey: Wiley-Interscience; 2006.

8. Stedman DH. Photochemical Ozone Formation, Simplified. Environ. Chem. 2004;1(2):65-6.

9. Melamed ML, Solomon S, Daniel JS, Langford AO, Portmann RW, Ryerson TB, et al. Measuring reactive nitrogen emissions from point sources using visible spectroscopy from aircraft. J. Environ.Monit. 2003;5:29-34.

10. Ibrahim O, Shaiganfar R, Sinreich R, Stein T, Platt U, Wagner T. Car MAX-DOAS measurements around entire cities: quantification of NOx emissions from the cities of Mannheim and Ludwigshafen (Germany). Atmos. Meas. Tech. 2010;3(3):709-21.

11. Wang S, Zhou B, Wang Z, Yang S, Hao N, Valks P, et al. Remote sensing of NO2 emission from the central urban area of Shanghai (China) using the mobile DOAS technique. J. Geophys. Res. 2012;117:D13305.

12. Beirle S, Boersma KF, Platt U, Lawrence MG, Wagner T. Megacity Emissions and Lifetimes of Nitrogen Oxides Probed from Space. Science. 2011;333(6050):1737-9.

13. Shaiganfar R, Beirle S, Sharma M, Chauhan A, Singh RP, Wagner T. Estimation of NOx emissions from Delhi using Car MAX-DOAS observations and comparison with OMI satellite data. Atmos. Chem. Phys. 2011;11(21):10871-87.

14. Kleinman LI, Daum PH, Imre D, Lee YN, Nunnermacker LJ, Springston SR, et al. Ozone production rate and hydrocarbon reactivity in 5 urban areas: A cause of high ozone concentration in Houston. Geophys.Res.Lett. 2002;29(10):1467.

15. Senff CJ, Alvarez RJ,II, Hardesty RM, Banta RM, Langford AO. Airborne lidar measurements of ozone flux downwind of Houston and Dallas. J. Geophys. Res. 2010;115:D20307.

16. Clapp L, Jenkin M. Analysis of the relationship between ambient levels Of O-3, NO2 and NO as a function of NOx in the UK. Atmos.Environ. 2001;35(36):6391-405.

17. Warneke C, de Gouw JA, Holloway JS, Peischl J, Ryerson TB, Atlas E, et al. Multi year trends in volatile organic compounds in Los Angeles, California: Five decades of decreasing emissions. J. Geophys. Res. 2012;117:D00V17.

18. Kim S.-W., Heckel A, Frost GJ, Richter A, Gleason J, Burrows JP, et al. NO2 columns in the western United States observed from space and simulated by a regional chemistry model and their implications for NOx emissions. J. Geophys. Res. 2009;114:D11301.

19. Langford AO, Brioude J, Cooper OR, Senff CJ, Alvarez RJ, II, Hardesty RM, et al. Stratospheric influence on surface ozone in the Los Angeles area during late spring and early summer of 2010. J. Geophys. Res. 2012;117:D00V06.

20. Langford AO, Senff CJ, Alvarez RJ,II, Banta RM, Hardesty RM. Long-range transport of ozone from the Los Angeles Basin: A case study. Geophys. Res. Lett. 2010;37:L06807.

21. Neuman JA, Trainer M, Aikin KC, Angevine WM, Brioude J, Brown SS, et al. Observations of ozone transport from the free troposphere to the Los Angeles basin. J. Geophys. Res. 2012;117:D00V09.

22.   Sillman S. The relation between ozone, NOx and hydrocarbons in urban and polluted rural environments.Atmos.Environ. 1999;33(12):1821-45.

23.   Liu SC, Trainer M, Fehsenfeld FC, Parrish DD, Williams EJ, Fahey DW, et al. Ozone Production in the Rural Troposphere and the Implications for Regional and Global Ozone Distributions. J. Geophys. Res. 1987;92(D4):4191-207.

24.   Lin X, Trainer M, Liu SC. On the Nonlinearity of the Tropospheric Ozone Production. J. Geophys. Res. 1988;93(D12):15879-88.

25.   Kleinman LI, Daum PH, Lee JH, Lee YN, Nunnermacker LJ, Springston SR, et al. Dependence of ozone production on NO and hydrocarbons in the troposphere. Geophys. Res. Lett. 1997;24(18):2299-302.

26.   Carter WPL, Lurmann FW. Evaluation of a Detailed Gas-Phase Atmospheric Reaction-Mechanism using Environmental Chamber Data. Atmos. Environ. 1991;25(12):2771-806.

27.   Carter WPL. Computer modeling of environmental chamber measurements of maximum incremental reactivities of volatile organic-compounds. Atmos. Environ. 1995;29(18):2513-27.

28.   Ren XR, Harder H, Martinez M, Lesher RL, Oliger A, Simpas JB, et al. OH and HO2 chemistry in the urban atmosphere of New York City. Atmos. Environ. 2003;37(26):3639-51.

29.   Volkamer R, Sheehy P, Molina LT, Molina MJ. Oxidative capacity of the Mexico City atmosphere - Part 1: A radical source perspective. Atmos. Chem. Phys. 2010;10(14):6969-91.

30.   Sheehy PM, Volkamer R, Molina LT, Molina MJ. Oxidative capacity of the Mexico City atmosphere - Part 2: A ROx radical cycling perspective. Atmos. Chem. Phys. 2010;10(14):6993-7008.

31.   Cazorla M, Brune WH, Ren X, Lefer B. Direct measurement of ozone production rates in Houston in 2009 and comparison with two estimation methods. Atmos. Chem. Phys. 2012;12(2):1203-12.

32.   Pusede SE, Cohen RC. On the observed response of ozone to NOx and VOC reactivity reductions in San Joaquin Valley California 1995-present. Atmos. Chem. Phys. 2012;12(18):8323-39.

33.   Richter A, Burrows JP, Nuss H, Granier C, Niemeier U. Increase in tropospheric nitrogen dioxide over China observed from space. Nature. 2005;437(7055):129-32.

34.   van der A RJ, Eskes HJ, Boersma KF, van Noije TPC, Van Roozendael M, De Smedt I, et al. Trends, seasonal variability and dominant NOx source derived from a ten year record of NO(2) measured from space. J. Geophys. Res. 2008;113(D4):D04302.

35.   Kim S-W, Heckel A, McKeen SA, Frost GJ, Hsie E-Y, Trainer MK, et al. Satellite-observed US power plant NOx emission reductions and their impact on air quality. Geophys. Res. Lett. 2006;33(22):L22812.

36.   Konovalov IB, Beekmann M, Burrows JP, Richter A. Satellite measurement based estimates of decadal changes in European nitrogen oxides emissions. Atmos. Chem. Phys. 2008;8(10):2623-41.

37.   Russell AR, Valin LC, Bucsela EJ, Wenig MO, Cohen RC. Space-based Constraints on Spatial and Temporal Patterns of NOx Emissions in California, 2005-2008. Environ. Sci. Technol. 2010;44(9):3608-15.

38.   Russell AR, Valin LC, Cohen RC. Trends in OMI NO2 observations over the United States: effects of emission control technology and the economic recession. Atmos. Chem. Phys. 2012;12(24):12197-209.

39.   Ryerson T, Andrews AE, Angevine WM, Bates TS, Brock CA, Cairns B, et al. The 2010 California research at the Nexus of air quality and climate change (CalNex) field study. J. Geophys. Res. 2013;118:5830-66.

40. Zaveri RA, Shaw WJ, Cziczo DJ, Schmid B, Ferrare RA, Alexander ML, et al. Overview of the 2010 Carbonaceous Aerosols and Radiative Effects Study (CARES). Atmos. Chem. Phys. 2012;12(16):7647-87.

41. Volkamer R, Coburn S, Dix B, Sinreich R. MAX-DOAS observations from ground, ship, and research aircraft: maximizing signal-to-noise to measure 'weak' absorbers. Proc. SPIE. 2009;7462-746203.

42. Baidar S, Oetjen H, Coburn S, Dix B, Ortega I, Sinreich R, et al. The CU Airborne MAX-DOAS instrument: vertical profiling of aerosol extinction and trace gases. Atmos. Meas. Tech. 2013;6(3):719-39.

43. Platt U, Stutz J. Differential Optical Absorption Spectroscopy: Principles and Applications. Heidelberg: Springer Verlag; 2008.

44. Oetjen H, Baidar S, Krotkov NA, Lamsal LN, Lechner M, Volkamer R. Airborne MAX-DOAS measurements over California: Testing the NASA OMI tropospheric NO2 product. J. Geophys. Res. Atmos. 2013;118:7400-13.

45. Alvarez RJ,II, Senff CJ, Langford AO, Weickmann AM, Law DC, Machol JL, et al. Development and Application of a Compact, Tunable, Solid-State Airborne Ozone Lidar System for Boundary Layer Profiling. J. Atmos. Ocean. Technol. 2011;28(10):1258-72.

46. Browell EV, Ismail S, Shipley ST. Ultraviolet Dial Measurements of O-3 Profiles in Regions of Spatially Inhomogeneous Aerosols. Appl. Opt. 1985;24(17):2827-36.

47. Davis KJ, Gamage N, Hagelberg CR, Kiemle C, Lenschow DH, Sullivan PP. An objective method for deriving atmospheric structure from airborne lidar observations. J. Atmos. Ocean. Technol. 2000;17(11):1455-68.

48. Langford AO, Senff CJ, Alvarez RJ, II, Banta RM, Hardesty RM, Parrish DD, et al. Comparison between the TOPAZ Airborne Ozone Lidar and In Situ Measurements during Tex AQS 2006. J. Atmos. Ocean. Technol. 2011;28(10):1243-57.

49. Pearson G, Davies F, Collier C. An analysis of the performance of the UFAM Pulsed Doppler Lidar for observing the boundary layer. J. Atmos. Ocean. Technol. 2009;26(2):240-50.

50. California Department of Conservation: GIS Mapping. 2013; Available at: http://www.conservation.ca.gov/dog/maps/Pages/GISMapping2.aspx. Accessed 09/25, 2013.

51. California Air Resource Board: 2010 CalNex Modeling Inventory. Available at: http://orthus.arb.ca.gov/calnex/data/. Accessed 09/25, 2013.

52. Hauglustaine DA, Granier C, Brasseur GP, Megie G. The Importance of Atmospheric Chemistry in the Calculation of Radiative Forcing on the Climate System. J. Geophys. Res. 1994;99(D1):1173-86.

2

Climate Change Adaptations for California's San Francisco Bay Area Water Supplies

William S. Sicke[1], Jay R. Lund[1*] and Josué Medellín-Azuara[1]

[1]*Department of Civil and Environmental Engineering, University of California, Davis, USA.*

Authors' contributions

This work was carried out in collaboration between all authors. All authors read and approved the final manuscript.

ABSTRACT

The impact of climate changes on both sea level and the temporal and spatial distribution of runoff will affect water supply reliability and operations in California. To meet future urban water demands in the San Francisco Bay Area, local water managers can adapt by changing water supply portfolios and operations. An engineering economic model, CALVIN, which optimizes water supply operations and allocations, was used to explore the effects on water supply of a severely warmer drier climate and substantial sea level rise, and to identify economically promising long-term adaptations for San Francisco Bay Area water systems. This modeling suggests that Bay Area urban water demands can be largely met, even under severe forms of climate change, but at a cost. Costs are from purchasing water from agricultural users (with agricultural opportunity costs), expensive water recycling and desalination alternatives, and some increases in water scarcity (costs of water conservation). The modeling also demonstrates the importance of water transfer and intertie infrastructure to facilitate flexible water management among Bay Area water agencies. The intertie capacity developed by Bay Area agencies for emergencies, such as earthquakes, becomes even more valuable for responding to severe changes in climate.

Keywords: Water supply; San Francisco Bay Area; engineering economic model; climate change; optimization.

---

*Corresponding author: Email: jrlund@ucdavis.edu;

1. INTRODUCTION

A changing climate will affect California's water supply management. The western United States and California can expect a shift in the temporal and spatial distribution of precipitation, changing streamflow, snowpack accumulation, snowmelt, and evapotranspiration [1,2,3]. These changes will affect the magnitude and timing of inflows into California's water supply system, affecting costs, operations, and allocations of water.

Higher average global temperature also will accelerate global sea level rise. Current projections suggest a mean sea level rise of 30 to 45 centimeters (cm) from year 2000 to 2050 [1]. Sea level rise and inland island failures will shift salinity of the Sacramento-San Joaquin Delta (Delta) inland and threaten the transfer of fresh water from northern California to the San Francisco Bay Area, San Joaquin and Tulare basins, and Southern California [4] [5] [6]. Sea level rise accompanied by a change in the Delta salinity could significantly affect the Delta as a major water supply hub [7].

Urban water management plans (UWMP) in California describe how water agencies plan to meet water demand under current hydrologic conditions and short-term and extended droughts. The California Department of Water Resources (DWR) requires updates to UWMPs every five years. In the San Francisco Bay Area, under current hydrologic conditions, urban water agencies rely on a portfolio of water sources including local inflows, groundwater (banking and pumping), water conservation, imported and transferred water, and water recycling. To mitigate potential shortages during droughts, the plans call for minimizing reliance on imported water through water conservation, expanded water recycling, desalination, firming up existing water transfer agreements, and entering into water transfer agreements [8,9,10,11,12,13,14,15,16,17]. East Bay Municipal Water District's (EBMUD) new investigation of expanding Contra Costa Water District's (CCWD) Los Vaqueros reservoir indicate the increasing sophistication of water planning in this region and the practical capability to respond to many future changes with considerable flexibility (albeit at some inconvenience and cost), as demonstrated by the results presented here.

This modeling effort explores potential effects of severe climate change on urban water supply in the San Francisco Bay Area and explores management actions to mitigate potential climate change impacts. This paper begins with an overview of the modeling approach used, including the climate change cases modeled. The next section presents and discusses the modeling results under several severe climate change cases, including water scarcity and the operating and scarcity costs, water supply portfolios, and infrastructure importance and expansion. The last section is a brief conclusion.

2. MODELING APPROACH

To better understand the local water management impacts from and adaptations to climate change in the context of statewide water supply management, a large scale economic-engineering optimization model, California Value Integrated Network (CALVIN), is employed. Such optimization modeling can identify promising qualitative management options, with details of system operations evaluated in later detailed simulation modeling.

2.1 CALVIN Model

CALVIN is an engineering optimization model of California's statewide intertied water supply system. Overall, CALVIN operates and allocates surface water and groundwater resources to minimize scarcity and operating costs, within the physical and environmental constraints of California's water supply system and selected policy constraints [18].

CALVIN has been used to explore various water management issues in California including conjunctive management of groundwater and surface water resources in Southern California, various forms of climate change, water markets in Southern California, and economic and water management effects of changes in Delta exports [19,20,21,22,23,24,25,26,27,28,29].

CALVIN is a generalized network flow model that uses the optimization solver Hydrologic Engineering Center's Prescriptive Reservoir Model (HEC-PRM) provided by the U.S. Army Corps of Engineers. CALVIN represents California's intertied water supply network, and includes 31 groundwater basins, 53 reservoirs, and 30 urban and 24 agricultural economically represented water demand areas (Fig. 1) covering 92 percent of California's population and 88 percent of its irrigated land.

CALVIN operates the physical infrastructure and allocates water within the system's constraints to minimize statewide water scarcity costs and operating costs. Scarcity occurs when an urban or agricultural delivery target is not met, and is defined as the difference between the target delivery (amount of water for which the user is willing to pay) and the volume of water delivered. Shortage (scarcity) costs are assigned to the unmet demand based on the user's economic willingness-to-pay (WTP) for additional water. Urban and agricultural water demand functions were scaled to 2050 population, with details of water demand estimation presented in Sicke et al. [30].

Equation 1 is the objective function used in CALVIN, and equations 2 through 4 are the constraints.

$$\text{Minimize:} \quad Z = \sum_i \sum_j c_{ij} X_{ij}, \tag{1}$$

$$\text{Subject to:} \quad \sum_i X_{ij} = \sum_i a_{ij} X_{ij} + b_j, \text{ for all nodes j}, \tag{2}$$

$$X_{ij} \le u_{ij} \quad \text{for all arcs}, \tag{3}$$

$$X_{ij} \ge l_{ij} \quad \text{for all arcs}, \tag{4}$$

where Z is the total cost of flows throughout the network, X_{ij} is flow volume leaving node i towards node j, c_{ij} = economic and operating unit costs (agricultural or urban), b_j = external inflow to node j, a_{ij} = gains/losses on flows in arc ij, u_{ij} = upper bound on arc ij, and l_{ij} = lower bound on arc ij.

Although the model does not explicitly include water quality, costs and constraints often are used to represent water quality considerations. Treatment costs for different water sources vary by their water quality and constraints limit the availability of some water sources due to their water quality or their ability to be blended. More detailed representation of water quality concerns are typically examined in later more detailed analyses. Many Bay Area utilities make considerable use of multiple water sources of differing qualities.

For each CALVIN optimization, model results include time series of monthly urban and agricultural water deliveries; stream, canal, and aqueduct flows; marginal value of additional water at every node in the network; the economic shadow values of the binding constraints; and the storage volumes in reservoirs and groundwater basins. Analysis and interpretation of these results provide insights into promising water management alternatives.

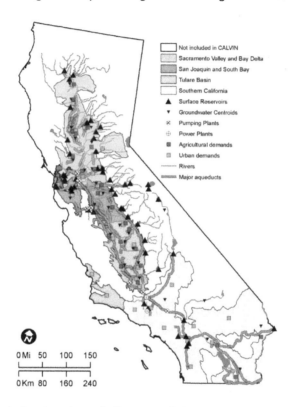

Fig. 1. Water supply infrastructure, inflows and demand areas represented in CALVIN

2.1.1 Operating costs

Operating costs in CALVIN include pumping, treatment, and water quality costs, as well as hydropower benefits (negative costs) [19]. Operating costs are all variable costs for pumping and treatment. The only exceptions are for additions of desalination and wastewater reuse capacity, where a linear average estimated total unit cost is used. CALVIN models most major facilities of California's intertied water supply system including recently completed Bay Area infrastructure such as the Freeport Regional Water Project (FRWP), the EBMUD-Hayward-San Francisco Public Utilities Commission (SFPUC) Intertie, and the EBMUD-CCWD Intertie. Urban areas were assumed to be able to recycle a portion of their wastewater flows (limited to local non-potable use). Urban areas with projected water recycling capacity by 2020 can use this as baseline recycling capacity at $500 per acre-ft. Urban areas with plans to expand water recycling capacity by 2050 were given expanded recycling capacity, up to 50 percent of urban wastewater flows, at $1,500 per acre-foot.

Additionally, urban coastal areas were allowed desalination at $2,100 per acre-foot. Earlier CALVIN water recycling and desalination costs are updated by Bartolomeo (E. Bartolomeo University of California Davis unpublished master's thesis 2011). Water recycling and desalination are capital intensive projects and ideally would be modeled as two-stage optimization with initial capital cost decisions and then operating costs decisions. In this study, we model total average annualized costs as operating costs. All costs are in 2008 dollars.

2.1.2 Bay area demand areas

Fig. 2a shows service areas for major water purveyors in the San Francisco Bay area. CALVIN aggregates purveyors into agricultural and urban demand areas. Aggregation is based on proximity and network connections, shown in Fig. 2b for the Bay Area.

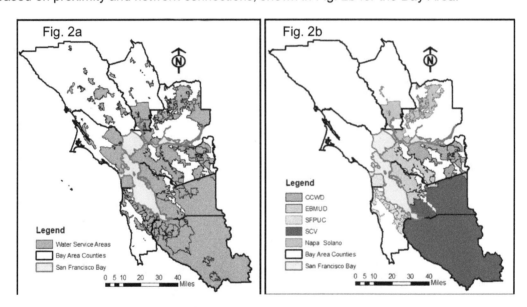

Fig. 2. Water supply retailers and wholesalers in the nine San Francisco bay area counties

Fig. 2a - water service areas boundaries. Fig. 2b - aggregated CALVIN urban demand areas.

2.1.3 Bay area supply sources and infrastructure

Five urban demand areas in CALVIN represent the San Francisco Bay Area portion of California's intertied water supply system. Each demand area has access to a variety of water sources. The overall statewide model schematic appears in Fig. 1, with the Bay Area network shown conceptually in Fig. 3. Seawater desalination is included as a potential future water source for all demand areas, except Napa-Solano, which is farthest upstream in the Delta.

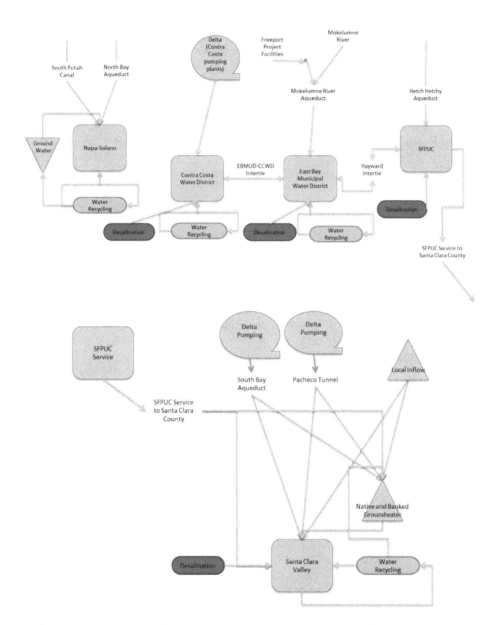

Fig. 3. Conceptualization of aggregate demand areas in the San Francisco bay area

Of the Bay Area demand areas, water for Napa-Solano is primarily from the United States Bureau of Reclamation's (USBR) Lake Berryessa, conveyed by the South Putah Canal, and State Water Project (SWP) water pumped from the northern Delta through the North Bay Aqueduct. Napa-Solano also uses small amounts of groundwater for Dixon and rural north Vacaville, and some water recycling [8,31].

Contra Costa Water District has its own water rights and also relies on USBR CVP water through pumping plants in the Delta (Mallard Slough, Rock Slough, and San Joaquin River). Other sources include water transfers along the EBMUD-CCWD Intertie and water recycling [10]. The EBMUD-CCWD Intertie was built for emergencies.

The East Bay Municipal Utility District relies primarily on imported water from the Mokelumne River Aqueduct, with storage at EBMUD's Pardee Reservoir. Some Sacramento River water can be conveyed through the recently completed USBR South Folsom Canal and Freeport Regional Water Project facilities. Other sources include water recycling and water transfers along the recently completed EBMUD-Hayward-SFPUC Intertie [11].

San Francisco Public Utilities Commission (SFPUC) relies principally on water from the Hetch Hetchy Reservoir on the Tuolumne River through the Hetch Hetchy Aqueduct. Other sources include water recycling and water transfers using the recently completed EBMUD-Hayward-SFPUC Intertie, and some local area inflows (omitted from the model due to data availability) [14].

In CALVIN, Santa Clara Valley water districts (SCV) include Santa Clara Valley Water District, Alameda County Water District, and Zone 7 Water Agency, the primary water suppliers of Alameda and Santa Clara counties. The SCV has access to a diverse water supply portfolio. SWP and CVP water is exported through Delta pumping and conveyed by the South Bay Aqueduct and San Luis Reservoir-Pacheco Tunnel respectively. SFPUC services the northern Santa Clara Valley to supplement water supply or to recharge groundwater. The SCV conjunctively uses surface and groundwater by banking local, imported, and recycled water in overdrafted aquifer space, giving it large naturally and artificially recharged groundwater supplies in the Livermore and Santa Clara Valleys. Other sources include water recycling [15,16,17].

Hydrologic variability is represented using 72 years of monthly hydrology data (1921–1993). Hydrologic representation includes surface and subsurface inflows and urban and agricultural return flows. Hydrologic inflows come from existing surface and integrated surface-groundwater models [18,19,32].

2.2 Modeling Climate Change

Two distinct climate changes are expected to affect water supply in California: changes in hydrology and sea level rise. Hydrologic change will be in the form of spatial and temporal distribution precipitation and streamflow. Sea level rise will affect salinity in the Sacramento-San Joaquin Delta [33]. Five severe future climate cases consider these changes: (1) a warm dry hydrology, (2) a historical hydrology and sea level rise that reduces Delta diversion capacity by 50 percent, (3) historical hydrology and sea level rise that ends Delta water diversions, (4) a warm dry hydrology and sea level rise that reduces Delta diversion capacity by 50 percent, and (5) both warm dry hydrology and sea level rise that ends Delta water diversions.

2.2.1 Hydrologic change

Global Circulation Models (GCMs) are often used to model climate change considering a range of emissions, population growth, socio-economic development, and technological progress. The Intergovernmental Panel on Climate Change (IPCC) Fourth Assessment Report 2007 describes these scenarios and summarizes temperature and precipitation climate projections. Regional GCM results and scenarios for California are discussed by Cayan et al. [2]. For input into CALVIN, the NOAA Geophysical Fluid Dynamics Laboratory CM2.1 A2 emissions scenarios were downscaled by Medellín-Azuara et al. [29,27] to capture the effects of a warm dry form of climate change by year 2050. The methods described in Zhu et al. [32] and Connell [28] were employed to perturb historical (1921–

1993) time series of inflows in CALVIN. Temperature and precipitation from the downscaled GFDL CM2.1 A2 scenario for a period of 30 years centered in year 2085 were used, indicating a 2°C (3.6°F) temperature increase and 3.5 percent decrease in precipitation.

Perturbation ratios for surface streamflows were built comparing a 30-year historical period centered in year 1979 with a future 30-year time period centered in year 2085. These were employed in Medellín-Azuara et al. [27] following the methods in Miller et al. [3]. Connell ([28] expanded the number of index rivers to 18, and showed that there were no significant gains in precision from adding more index river streamflows. This study employed the 18 index river information. Roughly a 27 percent statewide reduction in streamflows is expected under the GFDL CM2.1 A2 scenario. To perturb the 37 CALVIN rim inflows with the obtained 18 monthly river index ratios, correlation mapping was prepared following the methods in Zhu et al. [32] matching rim inflows with index rivers. Monthly time series of historical rim inflows in CALVIN were then multiplied by the ratio of the most correlated river index basin.

A linear relationship described in Zhu et al. [32] for each reservoir was used. Net evapotranspiration is obtained as the difference between evaporation and precipitation considering the area-elevation-capacity of each reservoir. Local accretions, on the other hand, use changes in deep percolation and precipitation in local areas.

2.2.2 Sea level rise

Most of the Delta is currently maintained as a largely fresh water system. This facilitates water movement from Northern California sources and storage to the Bay Area, southern Central Valley, and Southern California using pumping plants. The combined physical pumping capacities for the State Water Project (Banks), Central Valley Project (Jones), and the Contra Costa Water District (Old River and Rock Slough) are 16,500 cubic feet per second (cfs) (11.95 million acre-feet [maf]/year). Increasing sea level will increase Delta salinity to potentially reduce or end diversions from the Delta, by direct sea water intrusion, by collapsing some island levees which foster sea water intrusion, or a combination of both combined with stricter drinking water regulations [34]. Sea level rise is modeled in CALVIN by reducing the capacity of Delta pumping to 50 percent and to zero. This directly affects water users that rely on Delta water diversions.

2.2.3 Long-term urban water conservation

Long-term urban water conservation is implemented in the model to examine its value to reduce water supply related impacts of climate change (R. Ragatz University of California Davis unpublished master's thesis 2013). Thirty percent conservation in CALVIN results in average urban demand of 154 gpcd, similar to the State's 20 percent reduction goal by 2020 [35]. Costs to implement long-term urban water conservation are not addressed in this model, but would include outreach, public announcement campaigns, and efficient water use technologies.

2.2.4 Intertie conveyance policy constraints

Local San Francisco Bay Area water agencies have recently constructed interties to connect neighboring water agencies. Major interties are the EBMUD-Hayward-SFPUC Intertie, the EBMUD-CCWD Intertie, and the FRWP. Although the interties allow for the large water transfers, policy constraints can limit the frequency of use and available capacity.

The EBMUD-Hayward-SFPUC Intertie connects East Bay Municipal Utility District, San Francisco Public Utilities Commission, and the City of Hayward with a capacity of 30 million gallons per day (MGD) (Fig. 4). The EBMUD-CCWD Intertie connects East Bay Municipal Utility District and Contra Costa Water District with a capacity of 60 MGD to CCWD or 100 MGD to EBMUD (Fig. 4). Both the EBMUD-Hayward-SFPUC and the EBMUD-CCWD Interties were constructed for emergency response to catastrophic events such as an earthquake. The interties can boost water supply when needed and, under current policy agreements, are not intended for regular service.

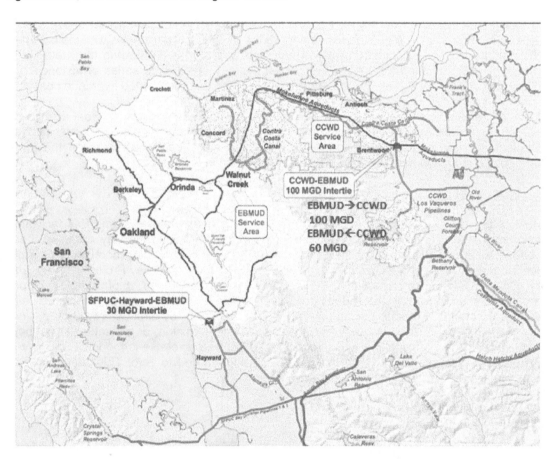

Fig. 4. Location and capacity of the EBMUD-hayward-SFPUC and EBMUD-CCWD Interties
(from EBMUD 2005 [11])

The Sacramento County Water Authority (SCWA) and East Bay Municipal Utility District joint project connects an intake at Freeport on the Sacramento River to the South Folsom Canal, and ultimately to EBMUD's Mokelumne River Aqueduct below Camanche Reservoir. The project can supply SCWA users with 85 MGD, with a 100 MGD intertie to EBMUD. Freeport operation is restricted to dry years or drought periods as defined by EBMUD's CVP contract. The effects of policy constraints on the interties were investigated by reducing the intertie capacities in the model to 20 percent of maximum capacity. The modeled unconstrained and constrained intertie capacities are listed in Table 1.

Table 1. Modeled intertie capacities (TAF = thousand acre-ft)

Intertie	Physical capacity TAF/Month	Policy constraint capacity TAF/Month
EBMUD-Hayward-SFPUC	2.8	0.56
EBMUD→CCWD	9.3	1.87
EBMUD←CCWD	5.6	1.12
Sacramento → EBMUD	9.3	1.87
SFPUC → Santa Clara Valley	13.5	2.7

2.2.5 Model runs

Eleven model runs were completed with CALVIN to evaluate three climate cases (Table 2). All runs use 2050 estimated population and land use. Specific urban land use responses to climate change are not included. Model run H is a base case for comparison with climate change scenarios, using historical hydrology to represent the spatial and temporal variability of inflows. Model runs WD, H-SLR50, H-SLR, WD-SLR50, and WD SLR represent the five climate change cases. Model run WD represents a warm dry future climate. Model run H-SLR50 represents a future climate where the hydrology is unchanged, but sea level rise reduces Delta diversion capacity by 50 percent. Model run H-SLR represents a future climate with unchanged hydrology, but sea level rise and other changes end major Delta water diversions. As modeled here, the sea level rise (combined with other Delta problems) is severe enough to significantly reduce or preclude all major Delta diversions [7]. Model run WD-SLR50 includes effects of a warm dry future climate combined with sea level rise that reduces Delta diversion capacity by 50 percent. Model run WD-SLR shows effects of a warm dry future climate combined with sea level rise that ends Delta diversions. Model runs "-C" explore effects of long-term urban water conservation on the impacts of climate changes. Model runs, "-P," evaluate the system flexibility gained by relaxing policy constraints on intertie operations. All climate change and policy constrained runs severely test the system, and do not individually represent likely futures.

The sea level rise cases that reduce Delta diversion capacity by 50 percent (-SLR50) do not show different average results from corresponding historical or the warm dry cases (e.g., H vs. H-SLR50 and WD vs. WD-SLR50), and so are not presented separately. This result is likely due to the significant storage available to the water supply network.

3. RESULTS

The results presented here, while preliminary, provide some perspective and insights on how the Bay Area could adapt to some fairly severe forms and consequences of climate change. Results from the model runs include water scarcity and scarcity costs, system operating costs, water supply portfolio, infrastructure importance and expansion opportunities, and environmental flow costs. Here we present a subset of these results. More detailed results on changing water supply portfolios, infrastructure importance and expansion opportunities, and environmental flow costs can be found in Sicke et al. [30].

3.1 Water Scarcity and Scarcity Cost

Under the climate change scenarios, water scarcity increases because of reduced inflows and reduced water diversions from the Delta. Scarcity costs represent the economic costs to water users from agricultural water shortages or costs of short-term conservation by households and businesses.

Table 2. Model runs

Run	Hydrology	Sea level rise	Long-term urban conservation	Intertie policy constraint
H (Base case)	Historical	None	None	None
Climate change				
H-SLR50	Historical	50% reduction	None	None
H-SLR	Historical	No Delta exports	None	None
WD	Warm Dry	None	None	None
WD-SLR50	Warm Dry	50% reduction	None	None
WD-SLR	Warm Dry	No Delta exports	None	None
Climate change and long-term urban water conservation				
H-SLR50-C	Historical	50% reduction	30% of Demand	None
H-SLR-C	Historical	No Delta exports	30% of Demand	None
WD-C	Warm Dry	None	30% of Demand	None
WD-SLR50-C	Warm Dry	50% reduction	30% of Demand	None
WD-SLR-C	Warm Dry	No Delta exports	30% of Demand	None
Climate change and policy constraints				
H-P	Historical	None	None	20% of Capacity
H-SLR50-P	Historical	50% reduction	None	20% of Capacity
H-SLR-P	Historical	No Delta exports	None	20% of Capacity
WD-P	Warm Dry	None	None	20% of Capacity
WD-SLR50-P	Warm Dry	50% reduction	None	20% of Capacity
WD-SLR-P	Warm Dry	No Delta exports	None	20% of Capacity

Table 3 shows scarcity, scarcity cost, and willingness-to-pay for additional deliveries for Bay Area urban water users and statewide urban and agricultural water users. Bay Area urban water demands are all met in the base case with the historical hydrology, although statewide users have annual average scarcity of 32 and 871 thousand acre-ft (taf) respectively. Water shortages in the base case reflect variability in water supply availability, infrastructure capacity, environmental flow constraints, and water supply costs that preclude some users from purchasing their full demand. Climate change impacts of reduced hydrology and sea level rise (no Delta exports) have little or no increased water scarcity for Bay Area water users, while statewide water users suffer more scarcity. Water shortages and shortage costs affect Santa Clara Valley districts the most under no export cases, as Santa Clara and Alameda counties rely on SWP and CVP imports. Table 3 shows that agricultural water users sell water and bear additional shortage costs of Bay Area and statewide urban users continuing to receive deliveries. Reduced water availability puts the agricultural sector in the position to sell water to the urban sector (spot, short-term or long-term transfers). The sea level rise case that reduces Delta diversion capacity by 50%s shows very small increases in scarcity or scarcity costs from the base case for the Bay Area. The scarcity and scarcity costs increase for agricultural demand. This result suggests there is an incentive for water transfers from agriculture to urban to meet urban demand.

Table 3. Average bay area urban water scarcity and scarcity cost

	Base case	Climate change			Climate change with long-term urban conservation			Historical hydrology and climate change with intertie policy Constraints			
	H	WD	H-SLR	WD-SLR	WD-C	H-SLR-C	WD-SLR-C	H-P	WD-P	H-SLR-P	WD-SLR-P
Scarcity, TAF/year											
Napa-Solano	0	0	0	0	0	0	0	0	0	0	0
CCWD	0	0	0	0	0	0	0	0	0	0	0
EBMUD	0	0	0	3	0	0	0	0	0	0	3
SFPUC	0	0	3	10	0	0	0	0	0	0	0
SCV-WD	0	0	26	26	0	0	0	0	0	26	26
Bay Area Urb	0	0	29	40	0	0	0	0	0	26	29
State Urb.	32	116	417	636	8	50	142	32	32	414	616
State Ag.	871	7,666	5,539	9,132	4,366	4,027	8,301	871	7,656	9,061	9,061
State Total	903	7,782	5,956	9,768	4,374	4,077	8,444	903	7,688	9,475	9,677
Scarcity Cost, $K/year											
Napa-Solano	0	0	0	0	0	0	0	0	0	0	0
CCWD	0	0	0	0	0	0	0	0	0	0	0
EBMUD	0	0	0	5,830	0	0	481	0	0	0	5,830
SFPUC	0	0	4,532	17,721	0	0	88	0	0	0	0
SCV-WD	0	0	46,495	46,495	0	0	0	0	0	46,495	46,495
Bay Area Urb	0	0	51,026	70,047	0	0	569	0	0	46,495	52,325
State Urb.	46,817	222,203	1,242,660	2,000,098	12,990	86,029	302,741	93,634	93,634	1,229,066	1,939,072
Average Marginal Willingness-to-Pay for additional water, $K/TAF											
Napa-Solano	0	0	0	0	0	0	0	0	0	224	224
CCWD	0	0	0	0	0	0	0	0	0	0	0
EBMUD	0	0	0	423	0	0	50	0	0	0	423
SFPUC	0	0	393	706	0	0	11	0	0	0	0
SCV-WD	0	0	751	751	0	0	0	0	0	751	751
Bay Area Urb	0	0	229	376	0	0	12	0	0	195	280
State Urban	25	86	263	420	23	52	106	25	25	241	378
State Ag.	33	230	186	301	148	162	285	33	230	299	299

H-Historical hydrology, WD-Warm dry hydrology, SLR-Sea level rise, C-Long-term urban Conservation, and P-Policy intertie constraints

The agricultural WTP for additional water listed in Table 3 is the average value of an additional unit of water to agricultural water users, representing the opportunity cost of transferring agricultural water to urban users. The agricultural opportunity cost is lowest under the no Delta diversions case and highest in the combined reduced streamflow and no Delta diversions case. Achieving 30 percent urban conservation alleviates all Bay Area urban shortages even under severe climate changes. Long-term urban water conservation also decreases economic motivation to transfer agricultural water to the urban sector.

Scarcity in the San Francisco Bay Area does not increase in the policy constraint model runs prohibiting Delta diversions. However, CCWD increases desalination to meet demand. With the EBMUD-CCWD Intertie constrained to 20 percent of its capacity, CCWD has no alterative water source when it cannot pump water from the Delta. Again, the scarcity and scarcity costs increase for agricultural demand. Urban demand in the Bay Area is met by agricultural to urban water transfers and increases in alternative water supply such as desalination. Operating costs include groundwater and surface water pumping, water treatment, waste water recycling, and desalination. Sicke et al. [30] show the average annual net operating variable costs. The most costly case is the combination of reduced streamflow and no Delta diversions. South of the Delta operating costs decrease because reduced water availability and no Delta exports reduces costly pumping. But other operating costs increase as urban water users turn to water recycling and desalination. Urban water conservation greatly reduces operating costs. Intertie policy constraints increase operating costs as expensive alternative water supply options such as much more expensive desalination are needed to meet high-value urban demand.

3.2 Supply Portfolios

Each Bay Area demand area relies on water supplies from a variety of local resources, imported water and water transfers, groundwater pumping, water recycling, and desalination sources. Climate change often shifts the mix of supplies as water becomes less available generally, as agricultural opportunity costs raise the cost of water transfers, and as water imports from the two large water projects (SWP and CVP) through the Delta are reduced. These factors can increase use of more costly options such as recycled water, desalination, groundwater banking, and pumping. Water supply portfolios add operational efficiency, and are often facilitated by a functional water market.

Santa Clara Valley Water District is the largest Bay Area urban demand area in CALVIN with a projected annual demand of 715,000 acre-ft by 2050. The SCV relies on imported SWP and CVP water from the Delta, local supplies, recycled water, and groundwater. Some of this area is supplied by SFPUC, represented as a water transfer to SCV in CALVIN. SCV also banks surface water in its aquifer. Fig. 5 and Sicke et al. [30] show how the supply portfolio for SCV shifts due to climate changes, long-term urban water conservation, and policy constraints on intertie operations.

Today, SCV relies heavily on SWP and CVP water from the Delta, averaging 253 thousand acre-feet/year (36 percent of water delivered). Groundwater pumping, local sources, and SFPUC account for about 17 percent each or 125 taf/year average. Water recycling averages about 2 percent of water supply, 16 taf/year. Water recycling has already reached capacity under the base case. A warm dry climate produces less local inflow, but increases Delta water imports slightly, suggesting water purchases from agricultural users to cover decreased local supplies. It is more economical to pay the agricultural opportunity cost than to expand desalination or wastewater reuse. The cases with a 50 percent reduction in Delta diversion capacity (SLR50) show little change in the water supply portfolio from the base case. The sea level rise cases (H-SLR and WD-SLR) have the largest effect on water supplies. When the Delta exports are unavailable, SCV can no longer rely on SWP and CVP water and water purchases from agricultural users become more restricted. Fig. 5 shows that in these cases scarcity and scarcity cost reach a point where SCV urban users are willing to pay to expand water recycling and desalination. Expanded water recycling capacity accounts for 18 percent of supply in both the sea level rise (H-SLR50) and warm dry and sea level rise (WD-SLR50) cases.

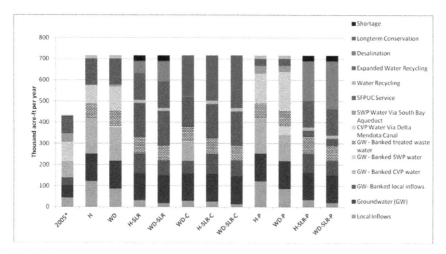

Fig.5. Average santa clara valley demand area water supply portfolio
*2005 water use estimates based on 2005 urban water management plans and State Water Plan
Abbreviations for model runs: H-Historical Hydrology, WD-Warm Dry Hydrology, SLR-Sea Level Rise, and C-Long-term Urban Water Conservation, and P -Policy Constraints on Interties

Desalination accounts for 9% and 14% of supply in the sea level rise (H-SLR50) and warm dry and sea level rise (WD-SLR50) cases, respectively. Water conservation of 30 percent in the warm dry climate case (WD-C) reduces dependence on imported SWP and CVP supplies. In the sea level rise (H-SLR50-C) and warm dry climate sea level rise cases (WD-SLR50-C), water conservation reduces the use of more expensive desalination/expanded wastewater reuse. The policy constraint model runs show that under sea level rise conditions, SCV must rely on high cost desalination in the absence of CVP and SWP supplies through the Delta and reduction in water from SFPUC. Under sea level rise, total groundwater banking/conjunctive use also drops.

Water management portfolios for other Bay Area demand areas are detailed in Sicke et al. [30]. The SFPUC demand area has access to water from the Hetch Hetchy Aqueduct, the Hayward Intertie, water recycling, desalination, and some local supplies. The policy constrained runs show that under all climate change the Hetch Hetchy supply is robust enough to maintain supply to SFPUC to meet 2050 demand. Small variation in the supply portfolio under unconstrained policy cases suggests operational cost savings through flexible operations of interties and water transfer agreements. SFPUC did not provide local inflow estimates to their system, so these were conservatively neglected in the model.

Contra Costa Water District demand area relies mainly on CVP water and its own water rights from the Delta conveyed through the Contra Costa Canal and Los Vaqueros Reservoir. In the base case model run, CVP water from the Delta is 92 percent of CCWD water supply, with the remainder from water recycling. In the warm dry climate, water recycling increases. Reducing Delta diversion capacity by 50% does not change the water supply portfolio. Ending Delta diversions affects the water supply portfolio differently, depending on the hydrology and intertie policy. With the historical hydrology, there may be sufficient water in the Mokelumne River system for the EBMUD-CCWD Intertie to offset the loss of Delta pumping. This would require purchasing water from other Mokelumne River diverters. Under limited water transfers through EBMUD-CCWD Intertie (H-SLR50-P and WD SLR50-P) desalination becomes cost effective in the absence of Delta diversions. In all warm dry climate cases combined with no Delta diversions, water recycling and desalination are expanded. Long-term urban water conservation reduces the need for costly desalination and water recycling.

The East Bay Municipal Utility District demand area relies mainly on water from the Mokelumne River Aqueduct. In the base case, Mokelumne River Aqueduct and transfers from CCWD account for all supplies [30]. Reduced Delta exports and diversion capacity do not significantly change the water supply portfolio. With ending Delta exports or diversions, water recycling makes up 3 percent of total water supply. With a warm dry climate, Freeport Project diversions become 31 percent of supply and water recycling expands to 9 percent. Sea level rise ends CCWD transfers of Delta water and reliance shifts heavily to Mokelumne River water. Combining a warm dry hydrology and sea level rise (ending Delta exports), EBMUD suffers small shortages on average, and must rely on all portfolio elements. With diversions from the Sacramento River north of the Delta reduced, EBMUD must rely on costly desalination.

The Napa-Solano demand area water supply portfolio base case and climate change cases are not significantly different. In all climate change cases, Napa-Solano relies on purchasing agricultural users' CVP and SWP water to respond to reductions in water availability. Napa-Solano demands are not affected by policy constraints on intertie operations. Being largely north of the Delta hydrologically eliminates problems from reduced south-of-the-Delta diversions.

3.3 Infrastructure Importance and Expansion Opportunities

CALVIN provides a platform for evaluating the importance of system components. The output from the network flow optimization solver provides shadow values (marginal value or cost) for each constraint in the model. Shadow values indicate the sensitivity of performance to capacity, flow, and availability uncertainty. The shadow value, in the case of storage capacity or conveyance capacity, represents the marginal value of that resource. The model runs for this analysis consider uncertain hydrology by looking at different climate change scenarios and looking at uncertainty and variability within each using the 72-year time series. Conveyance, water recycling, and storage capacities are represented in CALVIN as upper bounds on conveyance and storage links. Sensitivity output from CALVIN includes the marginal values of additional conveyance and storage capacity. When a conveyance or storage capacity is not reached in a time step, the marginal value of additional capacity is zero. The non-zero marginal value suggests a binding point in the system. A comparison of the marginal values between model runs suggests the importance of a system component, the relative flexibility of the system to manage climate change effects, and the potential for infrastructure expansion to improve system flexibility.

Table 4 contains the average value of one additional unit of increased capacity for selected conveyance and water recycling components in the Bay Area's water system.

Table 4. Average marginal value of conveyance and water recycling capacity ($/af)

Conveyance, Water recycling, and desalination infrastructure	Base case	Climate change						Climate change with long-term urban water conservation			Historical hydrology and climate change with intertie policy constraints			
	H	WD	H-SLR	WD-SLR	WD-C	H-SLR-C	WD-SLR-C	H-P	WD-P	H-SLR-P	WD-SLR-P
Freeport Project	0	44	4	1,122	25	0	379	0	446	5	1,135
Mokelumne River Aqueduct	0	0	114	15	0	0	0	0	0	0	0
Hetch Hetchy Aqueduct	204	137	535	414	7	39	11	1	2	1	1
CCWD-EBMUD Intertie	0	0	14	19	0	0	0	0	0	944	58
EBMUD-Hayward-SFPUC Intertie	160	150	518	104	46	76	176	141	494	138	932
SFPUC service to Santa Clara Valley	1	7	399	122	15	823	497	367	329	1,315	993
SCV Water Recycling	53	369	950	950	0	619	650	96	399	950	950
SCV Expanded Water Recycling	0	0	300	300	0	0	0	0	0	300	300
EMBUD Recycled water	0	88	1	956	0	0	240	0	501	0	927
EMBUD Expanded Recycled water	0	0	0	317	0	0	45	0	91	0	315
CCWD Water Recycling	7	264	310	1,280	124	100	458	2	256	1,050	1,050
CCWD Expanded Water Recycling	0	0	0	630	0	0	97	0	0	400	400

Abbreviations for model runs: H-Historical Hydrology, WD-Warm Dry Hydrology, SLR-Sea Level Rise, and C-Long-term Urban Water Conservation, and P-Policy Constraints on Interties

Fig. 6 shows the marginal values of infrastructure capacity expansion. For the Mokelumne River Aqueduct and the Freeport Project, on average, the capacity does not bind the system in the base case of historical hydrology. Under warm dry hydrologic conditions there is little to no change in the marginal values, because there is so little water in the system that the conveyance components rarely flow at capacity. The "-SLR50" cases results were omitted in the tables and figures because the results on average did not differ from the related historical and warm dry cases (e.g., H vs. H-SLR50 and WD vs. WD-SLR50).

The Hetch Hetchy Aqueduct constrains the system in all non-water conservation cases. This does not suggest that the Hetch Hetchy Aqueduct will not meet its primary design objective of supplying water to SFPUC. The Hetch Hetchy Aqueduct adequately meets demand under the policy-constrained intertie cases. These data suggest operational cost savings from intertie conveyance capacity. Ending exports or diversions from the Delta begins to stress the capacity of infrastructure as the model relies on conveyance through these remaining components to meet demand. Long-term urban water conservation reduces stress on these system components and reduces the value of increased capacity under climate change.

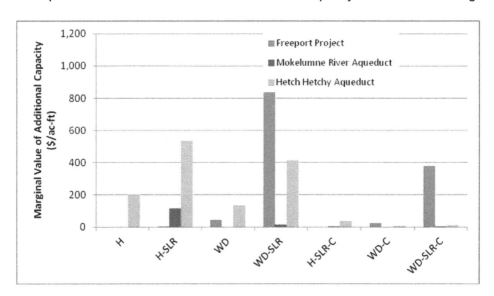

Fig. 6. Average marginal value of additional aqueduct conveyance capacity
Abbreviations for model runs: H-Historical Hydrology, WD-Warm Dry Hydrology, SLR-Sea Level Rise, and C-Long-term Urban Water Conservation

Fig. 7 shows the marginal value of intertie conveyance capacity. The CCWD-EBMUD Intertie's capacity is only slightly stressed on average under all climate changes. However, the EBMUD-Hayward-SFPUC Intertie increases in value under climate change conditions as more water users depend on the intertie to transfer and wheel water from various sources. Long-term urban water conservation reduces the marginal value of increased intertie capacity.

Fig. 8 shows the marginal value of base water recycling capacity and expanded water recycling capacity. Given the cost in CALVIN of base level water recycling at $500 per acre-ft and expanded water recycling at $1,500 per acre-ft, CALVIN uses base water recycling capacity before expanding water recycling capacity. The marginal values indicate the importance of water recycling capacity in increasing system flexibility for all climate changes. The marginal value of expanded water recycling capacity suggests the opportunity for infrastructure expansion mainly under a warm dry hydrology with reduced Delta supplies.

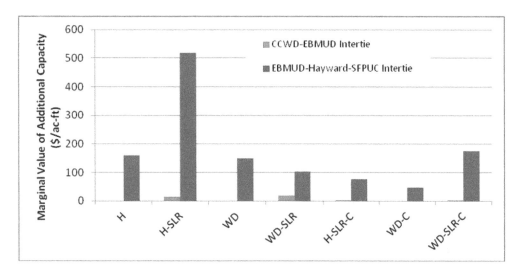

Fig. 7. Average marginal value of additional intertie conveyance capacity
Abbreviations for model runs: H-Historical Hydrology, WD-Warm Dry Hydrology, SLR-Sea Level Rise, and C-Long-term Urban Water Conservation

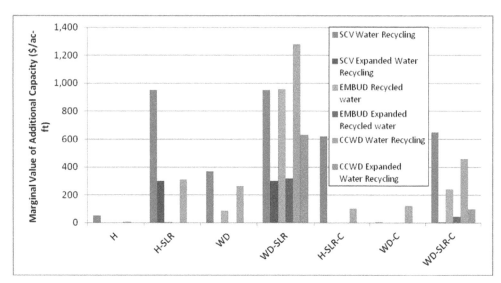

Fig. 8. Average marginal value of additional water recycling capacity
Abbreviations for model runs: H-Historical Hydrology, WD-Warm Dry Hydrology, SLR-Sea Level Rise, and C-Long-term Urban Water Conservation

The marginal value of increased storage capacity was surveyed over the entire system and generally showed that greater surface and groundwater storage capacity would not greatly increase the performance or flexibility of the water supply system (Table 5).

Table 5. Selected average marginal values of storage capacity ($/af)

Conveyance, water recycling, and desalination infrastructure	Base case	Climate change			Climate change with long-term urban water conservation			Historical hydrology and climate change with intertie policy constraints			
	H	WD	H-SLR	WD-SLR	WD-C	H-SLR-C	WD-SLR-C	H-P	WD-P	H-SLR-P	WD-SLR-P
Shasta Lake	5	45	5	23	35	5	23	6	53	2	14
Clair Engle Lake	2	27	2	20	21	2	22	5	47	3	12
Black Butte Lake	6	169	3	43	98	3	42	10	77	7	13
Lake Oroville	10	53	7	12	38	7	11	6	103	1	1
New Bullards Bar Reservoir	12	106	11	13	60	11	12	2	32	7	0
Folsom Lake	9	103	7	14	57	6	10	4	32	0	20
New Melones Reservoir	6	2	7	2	2	7	4	0	6	64	13
San Luis Reservoir	0	8	0	0	8	0	0	0	25	12	0
New Don Pedro Reservoir	6	3	6	2	3	6	4	3	3	11	0
Hetch Hetchy Reservoir	4	5	5	4	3	5	5	10	3	4	0

Abbreviations for model runs: H-Historical Hydrology, WD-Warm Dry Hydrology, SLR-Sea Level Rise, and C-Long-term Urban Water Conservation, and P-Policy Constraints on Interties

However, the robust existing water storage capacity raised system resiliency in the "-SLR50" cases (sea level rise cases that model reduced Delta diversion capacity by 50 percent). The "-SLR50" cases results were excluded here because the results on average did not differ from the related historical and warm dry cases (e.g., H vs. H-SLR50 and WD vs. WD-SLR50). Future work could include model cases that look at the performance of the reservoir systems in managing seasonal changes Delta salinity that could seasonally affect diversion capacity.

Fig. 9 shows marginal values for relaxing policy constraints on intertie operations. The interties increase the variety of an agency's water supply portfolio, allow for wheeling of water among agencies, and allow agencies to cooperate on alternatives such as water recycling and desalination. The CCWD-EBMUD intertie is very important for reducing shortages at CCWD. Overall, the interties are most valuable with Delta diversion restrictions.

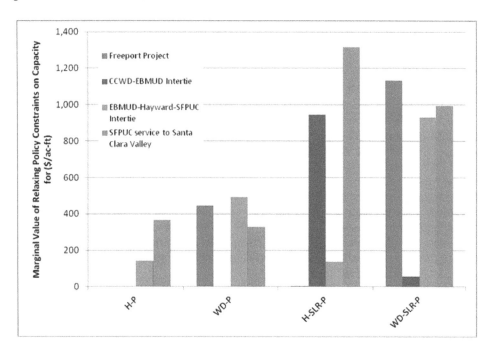

Fig. 9. Average marginal value of relaxing policy constraints on intertie capacity

4. LIMITATIONS

CALVIN, like all models, is merely a representation of a real system and suffers from limitations. A comprehensive list of CALVIN limitations are discussed by Jenkins et al. [19] [23]. For this particular study, CALVIN limitations are discussed in more depth in Sicke et al. [30]. These limitations include: water demands and efficiencies do not vary by water year type, other uncertainties in water demands arising from urban water conservation efforts and their costs, long-term changes in environmental requirements, the hydrologic foresight of the modeled operations, the pricing of water transfer agreements, localized water distribution system constraints, and operations occurring on a time scale less than one month.

CALVIN results can be improved with updates from the forthcoming 2010 Urban Water Management Plan data, particularly regarding base water demands in the Bay Area. Despite the limitations described, this reconnaissance level modeling analysis highlights many opportunities for the Bay Area's complex water system to adapt to fairly extreme forms of climate change.

5. CONCLUSIONS

The San Francisco Bay Area has the economic and infrastructure potential to weather quite severe forms of climate change, at some costs and assuming operational flexibility by Bay Area water providers and regulators. This adaptation potential is largely made possible by a series of system interties completed in recent years for emergency response purposes, but which also can provide longer-term benefits and flexibility.

Water markets allow urban water users in the Bay Area to operate flexibly and purchase water from agricultural users and each other. The SFPUC and EBMUD, with their access to Hetch Hetchy Aqueduct and Mokelumne River Aqueduct water, rely less on the Delta but may see economic benefit from water recycling and desalination under unfavorable climate changes. The SFPUC and EBMUD are not necessarily turning to alternative water supplies because of reduced Hetch Hetchy or Mokelumne River Aqueduct water. The EBMUD-Hayward-SFPUC and EBMUD-CCWD Interties combined with SFPUC service in Santa Clara Valley allows for purchases and transfers of imported water (Hetch Hetchy and Mokelumne River Aqueducts), recycled water, and desalination water to the demand areas that have lost access to CVP and SWP water or suffered reduced regional inflows, thus providing operating and scarcity cost savings.

Water recycling and desalination also can improve water supply reliability by reducing reliance of imported water supply. Under fairly severe climate change conditions, especially with sea level rise ending water diversions from the Sacramento-San Joaquin Delta, purchasing agricultural water becomes more expensive. The CVP and SWP water purchases and transfers wheeled through the Delta become restricted, and urban water users turn to more costly water supply alternatives such as water recycling and desalination, affecting SCV and CCWD the most.

Long-term urban water conservation greatly decreases the effects of severe climate change, and reduces operating costs and reliance on expensive supply alternatives such as water recycling and desalination. However, expanding water conservation will require extensive planning and some costs.

Overall, adaptation to a warmer drier climate relies primarily on improved system flexibility with investments in water recycling and desalination, at a cost, while adaptation to ending Delta diversions relies on alternative water supply and water transfers along the existing emergency interties which are important to system flexibility. Challenges to water management will be policies, agreements, and regulations that allow for flexible water transfers, more than mere existence of physical infrastructure. The average yearly cost for the intertie policy constraints were $51 million, $297 million, and $896 million for the warm dry, no Delta diversion, and warm dry hydrology with no Delta diversion model runs, respectively. A management policy for intertie cooperative operations can allow large investments in water recycling and desalination to be shared by several agencies.

ACKNOWLEDGEMENTS

The authors thank the following individuals for their help in preparing, reviewing, and improving this work: Christina Buck, Rachel Ragatz, Guido Franco, Sarah Pittiglio, Michael Hanemann, Rebecca Chaplin-Kramer, David Behar, Clifford Chan, Benjamin Bray, Brian Campbell, Raffi J. Moughamian, Maureen Martin, Leah Orloff, Sarah Young, John Andrew, Dan Cayan, Scripps, Susi Moser, Bruce Riordan, and Mark Wilson. Of course, all remaining shortcomings are due to the authors.

COMPETING INTERESTS

Authors have declared that no competing interests exist.

REFERENCES

1. Cayan D, Tyree R, Dettinger M, Hidalgo H, Das T, Maurer E, Bromirski P, Graham M, Flick R. Climate Change Scenarios and Sea Level Rise Estimates for the California 2009, Climate Change Scenarios Assessment. California Climate Change Center, Publication NumberCEC-500-2009-014-F. 2009;64.
2. Cayan D, Maurer E, Dettinger M, Tyree M, Hayhoe K. Climate change scenarios for the California region. Climate Change. 2008;87:S21–S42.
3. Miller N, Bashford K, Strem E. "Potential impacts of climate change on California hydrology." Journal American Water Works Association. 2008;39:771–784.
4. Fleenor W, Hanak E, Lund J, Mount J. Delta hydrodynamics and water quality with future conditions. Appendix C to Lund J. et al. Comparing futures for the Sacramento-San Joaquin Delta. San Francisco: Public Policy Institute of California; 2008.
5. William P. An Overview of the Impact of Accelerated Sea Level Rise on San Francisco Bay. 1985: Project No. 256, for the Bay Conservation and Development Commission. Philip Williams and Associates, San Francisco, California.
6. Williams PB. The Impacts of Climate Change on the Salinity of San Francisco Bay. Philip Williams and Associates, San Francisco, California. EPA Report # EPA-230-05-89-051; 1987.
7. Lund J, Hanak E, Fleenor W, Bennett W, Howitt R, Mount J, Moyle P. Comparing Futures for the Sacramento-San Joaquin Delta, University of California Press, Berkeley, California; 2010.
8. City of Napa. Urban Water Management Plan. (Napa UWMP); 2005.
9. City of Sonoma. Urban Water Management Plan. (Sonoma UWMP); 2005.
10. Contra Costa Water District. Urban Water Management Plan. (CCWD UWMP); 2005.
11. East Bay Municipal Utility District. Urban Water Management Plan. (EBMUD UWMP); 2005.
12. Marin Municipal Water District. Urban Water Management Plan. (Marin UWMP); 2005.
13. North Marin Water District. Urban Water Management Plan. (North Marin UWMP); 2005.
14. San Francisco Public Utility Commission. Urban Water Management Plan; (SFPUC UWMP); 2005
15. Santa Clara Valley Water District. Urban Water Management Plan. (SCVWD UWMP); 2005.
16. Zone 7 Water Agency. Urban Water Management Plan. (Zone 7 UWMP); 2005.
17. Alameda County. Urban Water Management Plan. (Alameda UWMP); 2005.

18. Draper J, Jenkins MW, Kirby K, Lund J, Howitt R. Economic- Engineering Optimization for California Water Management. Journal of Water Resources Planning and Management. 2003;129(3):155–164.

19. Jenkins M, Draper A, Lund J, Howitt R, Tanaka S, Ritzema R, Marques G, Msangi S, Newlin B, Van Lienden B, Davis M, Ward K. Improving California Water Management: Optimizing Value and Flexibility. Center for Environmental and Water Resources Engineering Report No. 01-1, Dept. of Civil and Environmental Engineering, University of California, Davis, California; 2001.

20. Newlin B, Jenkins M, Lund J, Howitt R. Southern California Water Markets: Potential and Limitations. Journal of Water Resources Planning and Management; 2002.

21. Tanaka S, Lund J, Jenkins M. Effects of Increased Delta Exports on Sacramento Valley's Economy and Water Management." Journal of the American Water Resources Association. 2003;39(6):1509–1519.

22. Tanaka S, Zhu T, Lund J, Howitt R, Jenkins M, Pulido M, Tauber M, Ritzema R, Ferreira I. Climate Warming and Water Management Adaptation for California. Climatic Change. 2006;76(3-4):361–387.

23. Jenkins M, Lund J, Howitt R, Draper A, Msangi S, Tanaka S, Ritzema R, Marques G. Optimization of California's Water System: Results and Insights. Journal of Water Resources Planning and Management. 2004;130(4).

24. Lund J, Hanak E, Fleenor W, Howitt R, Mount J, Moyle P. Envisioning Futures for the Sacramento-San Joaquin Delta. Public Policy Institute of California, San Francisco, California; 2007.

25. Null S, Lund J. Re-Assembling Hetch Hetchy: Water Supply Implications of Removing O'Shaughnessy Dam. J. of the American Water Resources Association. 2006;42(2):395–408.

26. Pulido-Velázquez M, Jenkins M, Lund J. Economic Values for Conjunctive Use and Water Banking in Southern California. Water Resources Research. 2004;40(3).

27. Medellín-Azuara J, Harou J, Olivares M, Madani K, Lund J, Howitt R, Tanaka S, Jenkins M, Zhu T. Adaptability and Adaptations of California's Water Supply System to Dry Climate Warming. Climatic Change. 2008;87(Suppl 1):S75–S90.

28. Connell-Buck C. Adapting California's Water System to Warm vs. Dry Climates." Climatic Change. 2011.109(Suppl 1):S133-149.

29. Medellín-Azuara J, Connell C, Madani K, Lund J, Howitt R. Water Management Adaptation with Climate Change, California Energy Commission, Public Interest Energy Research (PIER); 2009, Sacramento, California. p. 30. Available at http://www.energy.ca.gov/2009publications/CEC-500-2009-049/CEC-500-2009-049-F.PDF. Accessed 01 November 2009.

30. Sicke W, Lund J, Medellín-Azuara J. (UC Davis Department of Civil and Environmental Engineering). 2012. Climate Change Adaptations for Local Water Management in the San Francisco Bay Area. California Energy Commission. Publication number: CEC-500-2012-XXX.

31. Cal Water Dixon. Urban Water Management Plan (Cal Water Dixon UWMP); 2005.

32. Zhu T, Jenkins M, Lund J. Estimated impacts of climate warming on California water availability under twelve future climate scenarios. Journal of the American Water Resources Association. 2005;41:1027–1038.

33. DWR. California Climate Adaptation Strategy. California Department of Water Resources. A Report to the Governor of the State Of California; 2009.

34. Chen W, Haunschild K, Lund J, Fleenor W. Current and Long-Term Effects of Delta Water Quality on Drinking Water Treatment Costs from Disinfection Byproduct Formation. San Francisco Estuary and Watershed Science. 2010;8(3).

35. SWRCB. (State Water Resources Control Board). 2010 20x2020 Water Conservation Plan; 2010.

3

Assessment of Municipal Effluent Reclamation Process Based on the Information of Cost Analysis and Environmental Impacts

Yu-De Huang[1], Hsin-Hsu Huang[1], Ching-Ping Chu[1*] and Yu-Jen Chung[1]

[1]Environmental Engineering Research Center, Sinotech Engineering Consultants, Inc., 6F, No.280, Xin Hu 2nd Rd., Nei Hu Dist, Taipei City, Taiwan.

Authors' contributions

This work was carried out in collaboration between all authors. All authors read and approved the final manuscript.

ABSTRACT

Water shortage has now become a global issue. Reclamation of the effluent from municipal wastewater treatment plant is feasible for supplying the quick growth of water requirement. The objective of this study was to conduct both the cost analysis and environmental impact evaluation of two reclamation processes: sand filter – ultrafiltration - reverse osmosis (SF-UF-RO) and sand filter - electrodialysis reversal (SF-EDR). The results will serve as a reference for selecting the process in the scale-up construction works. Two processes were installed in a reclamation pilot plant in Futian Water Resource Recycling Center (Taichung City, Taiwan) and operated in parallel to evaluate their stability and product quality. The cost analysis was conducted to estimate the capital requirement of building large-scale plant for reclaiming the effluent. The cost of land construction, mechanical with electronic equipment and operation with maintenance were all considered in the analysis. On the other hand, the environmental assessment of these processes has been realized by Life Cycle Assessment (LCA). The software Sima Pro 7.3 was used as the LCA analysis tool. Four different evaluation methods, including Eco-indicator 99, Ecopoints 97, Impact 2002+ and CML 2 baseline 2000, were applied. The results show that the water quality of SF-EDR has similar potential in reclaiming the effluent from municipal water resource recycling center as SF-UF-RO. The cost of SF-EDR is lower than that of SF-UF-RO. In the environmental analysis, the LCA demonstrates that SF-EDR may create more impacts on the environment due to more consumption on electricity and chemicals than SF-UF-RO. Using SF-UF-RO as the effluent reclamation process may be an option causing less impacts on climate change.

Keywords: Reverse osmosis (RO); Electrodialysis reversal (EDR); municipal wastewater reclamation; cost analysis; environmental impact comparison.

1. INTRODUCTION

In the recent decades, water consumption in Taiwan has gradually increased due to population growth, urbanization and the rapid development of industries. However, the construction of reservoirs and groundwater wells has become increasingly difficult because of the strong public concerns. Reclamation of effluent from municipal wastewater treatment plant has been considered as a feasible solution as the effluent quality is usually stable and acceptable [1,2]. The reclaimed effluent can then be used to supply the requirement, especially for those from industries with enormous water consumption [3,4]. Successful experiences of the large cases also show that this solution is feasible, such as "NEWater" in Singapore (microfiltration-reverse osmosis-ultraviolet), "Groundwater Replenishment System" in Orange County, California, USA (microfiltration-reverse osmosis-advanced oxidation), and "Water Reclamation and Management Scheme" in Sydney, Australia (microfiltration-reverse osmosis-chlorine disinfection). Furthermore, the quick growth of public sewer construction and connection in urban areas of Taiwan provides an appropriate environment to promote this idea.

In the previous cases, the combination of ultrafiltration (UF) and reserve osmosis (RO) is a widely accepted process to reclaim the effluent from wastewater treatment plants. This is called the "dual membrane process". In Taiwan, a variety of local studies also showed its feasibility and the performance stability [3,4]. On the other hand, electrodialysis reversal (EDR) has been less applied in this field. Due to the structure of flowing channels, it requires less pre-treatment than that of RO. Compared to RO, EDR has a tolerance to colloids, microorganisms and silicate, and thus it allows the SDI of inlet up to 12, while the SDI requirement of RO inlet is less 3. The total cost of using EDR to desalinate the municipal wastewater may be reduced. In UF-RO process, the permeate flow is driven by pressure drop across the membrane. For EDR, electric current forces dissolved salt ions in the wastewater through an electrodialysis stack consisting of alternating layers of cationic and anionic ion exchange membranes. The direction of ion flow is periodically reversed by reversing the polarity of applied electric current to prevent the scaling on the membranes. More power consumption is possibly required for EDR than UF-RO. For the water quality, UF-RO usually gives better permeate than EDR, especially in turbidity, total organic carbon and conductivity. The removal rate of the multivalent ions is worse in the case of EDR than RO [5]. In case that the requirement of production conductivity is less stringent, EDR is a suitable application, such as replenishing water in the recirculation cooling system [6].

In general, capital investment and operation cost is the main factor for choosing reclamation process. From some other angles, in Taiwan, carbon emission inventory has gradually become a necessary step in the planning stage. Other environmental impacts induced by the electricity and chemicals, including the greenhouse effects and other pollutions, should also be considered in the evaluation from the aspect of whole life cycle analysis (LCA). LCA is a method to study the environmental aspects and potential impacts throughout a product's life cycle starting from raw material acquisition, manufacture, use, recycling and disposal. LCA is helpful in measuring the ecological aspects of products composed of different raw materials though used for the same purposes. In 1990s, The Society of Environment Toxicology and Chemistry (SETAC) has drafted a series of handbooks for conducting LCA. In June 1997, "ISO 14040 Life cycle assessment - Principles and framework" has been announced. After that, LCA is more widely applied in the environmental management strategies of enterprises. When selecting environmentally friendly raw materials, products, and production processes, the results of a comparative LCA study provide reference value for decision makers [7].

Sima Pro (PRé Consultants bv, the Netherlands) is a widely adopted LCA software based on ISO 14040 and has been applied in the decision making. It was first developed by Leiden University, the Netherlands in 1990 and includes a variety of databases. Using the information in the database, the input of raw materials and output of pollutants can be transformed into the impacts on the ecology and environment. Using the function of normalization, the impacts of different processes or using different raw materials can be quantified and compared.

In many cases for selecting proper water or wastewater treatment processes, Sima Pro has been applied to analyze the environmental impacts raised from the consumption of chemicals and electricity, as well as the waste generation. The assessment usually focuses on the fields such as greenhouse effect, resources depletion, acidification, eutrophication and human toxicity [8]. Ortiz et al. [9] conducted environmental analysis on the conventional activated sludge process and the additional tertiary treatment units, and evaluated whether the additional tertiary units increased the environmental impacts. Bonton et al. [10] conducted a comparative life cycle assessment of two surface water treatment plants, using enhanced conventional process and nanofiltration membrane, respectively. Barrios et al. [11] compared the financial cost and environmental costs of individual units in a water treatment plant, and quantitatively assessed if any more environmentally friendly alternative could be applied. Muñoz et al. [12] compared the environmental impacts between the homogeneous and heterogeneous advanced oxidation processes (AOP) on treating industrial wastewater. Nijdam et al. [13] evaluated the impacts between two processes, granular activated carbon (GAC) and ozone/UV, on treating leachate from the landfill. Chatzisymeon et al. [14] assessed the environmental analysis among three AOPs, UV heterogenous photocatalysis (UV/TiO$_2$), wet air oxidation (WAO) and electrochemical oxidation (EO), on treating olive mill wastewater. The main difference of these processes not only came from the electricity consumption, but also the catalyst amount used during the treatment. In these studies, multiple evaluating methods are applied so that we may give credits to the individual process from different aspects. Based on the results, it can give a full picture regarding which process is more suitable and impacts the environment less.

Using desalination units like RO or EDR to produce clean water from the discharged effluent certainly consumes more energy than the traditional water treatment plants. When the scale of effluent reclamation plant gradually becomes larger since the last decade, say, over 100,000 m^3 per day, it raises more concerns regarding the greenhouse gas emission. In the literature, to the best of authors' knowledge, the studies on EDR and RO mainly focused on the production quality and the cost [6,15,16]. The comprehensive analysis on the environmental impacts of the two processes however is still lack. In this study, we conducted both the cost analysis and environmental impact assessment of two reclamation processes: SF-UF-RO (sand filter – ultrafiltration - reverse osmosis) and SF-EDR (sand filter-electrodialysis reversal). A variety of environmental impacts, including global warning, human toxicity, eutrophication, acidification, ozone layer depletion and so on has been considered. The results would serve as a reference for the decision making for the full-scale construction of an effluent reclamation plant in the near future.

2. MATERIALS AND METHODS

2.1 Futian Municipal Wastewater Treatment Plant

In this study, the chosen wastewater treatment plant is Futian Water Resource Recycling Center, located in Taichung, a large city in central Taiwan. Quick growth of industries in this region is the main reason leading to its water shortage. To relieve the shortage, Futian Water Resource Recycling Center has been considered as a feasible site for upgrading and reclaiming the effluent for industrial use (Fig. 1). The effluent is of good quality and the reclaimed effluent can be supplied to the steel-making industries as coolant water.

Fig. 1. The location of Futian water resource recycling center and the industrial parks in Taichung City

Futian Water Resource Recycling Center is located in the southern district of Taichung City, which currently applies activated sludge process to treat 55,000 m^3 of sewage daily (the first phase). The footprint of this plant is 13.6 hectares. After being treated by secondary clarifier and chlorine disinfection, the effluent has been discharged to the Green Stream in the neighborhood. The sludge is treated by gravitational thickening, anaerobic digestion, and belt dewatering. In year 2030, the capacity will increase to 295,000 CMD in its fourth (final) phase.

To evaluate the feasibility to reclaim the effluent, a full survey has been commenced since 2006. In its effluent, the suspended solids was 5.5 mg/L on average, biochemical oxygen demand was 2.4 mg/L on average, chemical oxygen demand was 10 mg/L on average, and pH ranged from 7.0 to 7.5. For the salty items, the hardness ranged from 100 to 170 mg/L, and the total dissolved solids (TDS) ranged from 250 to 350 mg/L. The effluent quality is acceptable for many industrial uses if it is further purified.

2.2 Pilot Plant for Effluent Reclamation

To evaluate the feasibility and to select the proper process, a pilot plant has been installed in Futian Water Resource Recycling Center to continuously reclaim this effluent. It included two processes, "Sand Filter - Electrodialysis Reversal" (SF-EDR) and "Sand Filter - Ultrafiltration - Reverse Osmosis" (SF-UF-RO) in parallel (Fig. 2). The pilot plant installed in Futian Water Resource Recycling Center continuously reclaims this effluent using the two processes. The operation began in October 2008. The performances of the two processes were compared to reveal if they were efficient in reclaiming the effluent, how much they cost, and how stable they performed. Table 1 compares the design parameters and the actual parameters after long-term operation.

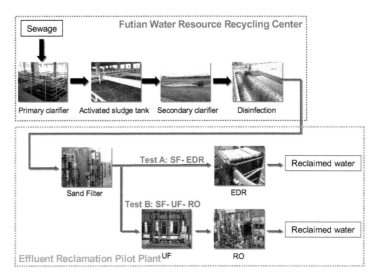

Fig. 2. The process of effluent reclamation pilot plant (SF-UF-RO and SF-EDR)

Table 1. The design and operation parameters of sf, uf, ro and edr in the pilot plant

Process Unit	Parameters	Operation Values
Sand filter	Capacity (CMD)	120
	Flux (m^3/m^2-hr)	8.68
Ultrafiltration	Capacity (CMD)	120
	Flux (m^3/m^2-hr)	0.072
	SDI15 of product	1~1.5
Reverse osmosis	Capacity (CMD)	60
	Flux (m^3/m^2-hr)	0.037
	Recovery (%)	55
	Desalination efficiency (%)	90~94%
Electrodialysis reversal	Capacity (CMD)	4
	Desalination efficiency (%)	80
	Recovery (%)	50
	Operation voltage (V)	84
	Circulation rate (L/min)	18
	Electrode reversal period (hr)	1

2.3 Evaluation I: Cost Analysis

The detailed items of cost analysis were composed of three parts [16]:

1. Land construction (L), including the plant building for installing the relevant facilities, piping systems and storage tanks.
2. Mechanical and electronic equipments (M), including the membrane modules, pumps, blowers, and the monitoring meters.
3. Operation and maintenance (K), including the consumables (chemicals, prefilters and membrane elements), electricity consumption, personnel expenses and the insurance during the operation.

One may obtain the cost to produce one ton of filtrate from the two processes of filtrate production scale Q (in unit of m^3/day), denoted as c, using the following equations:

$$c = \frac{Y + K}{Q * 365} \qquad (1)$$

$$Y = (L + M) \times \left\{ (1+r) \left[i + \frac{j}{(1+j)^t - 1} + x \right] + f \right\} \qquad (2)$$

The parameters are defined as:

r: the safety factor of public works budget in Taiwan, 47% in this study
t: the depreciation period, 20 years in this study
i: annual loan interest rate, 6% in this study
j: annual deposit interest rate, 3% in this study
x: the tax rate and relevant insurances during the construction, 0.62% in this study
f: percentage of annual facility renewal, 1.36% in this study

2.4 Evaluation II: Environmental Impact Assessment

Comparisons of environmental impacts of two processes were performed using the software SimaPro 7.3 (the registration name of SimaPro 7.3 which used in this study is Sinotech). The environmental impacts of reclamation processes are intrinsically related to its direct discharge of pollutants in the retentate (from RO or EDR), electricity consumption, consumables (like membrane, prefilters and chemicals), the piping works of reclaimed water, and disposal of waste. In order to quantify the aforementioned environmental impacts of different process as possible, a variety of methods have been applied in this study. The four methods, Eco-indicator 99, Ecopoints 97, Impact 2002+, and CML 2 baseline 2000, mainly consider the overall impacts on the ecology and human health. Table 2 lists the characteristics of the four methods.

Table 2. Characteristics of methods used in this study

Method	Introduction	Impact Categories
Eco-indicator 99	The damage-oriented approach. It is to assess the seriousness of three damage categories: 1. damage to human health; 2. damage to ecosystem; 3. damage to resources.	Carcinogens, respiratory organics, respiratory inorganics, climate change, radiation, ozone layer, ecotoxicity, acidification/ eutrophication, land use, minerals
Ecopoints 97	Based on actual pollution and critical targets that are derived from Swiss policy. The following data are necessary in calculating a score in ecopoints for a given product: 1. quantified impacts of a product; 2. total environmental load for each impact type in a particular geographical area; 3. maximum acceptable environmental load for each impact type in that particular geographical area.	It assesses impacts, such as air pollution (NOx, SOx, PM10, etc.) , water pollution (COD, P, heavy metals, etc.) and solid waste, individually
Impact 2002+	Proposes a feasible implementation of a combined midpoint/damage approach, linking all types of life cycle inventory results (elementary flows and other interventions) via 14 midpoint categories to four damage categories.	Ionizing radiation, Carcinogens, Non-carcinogens, Respiratory organics, Ozone layer depletion, Global warming, Respiratory inorganics, Aquatic eutrophication, Terrestrial acid/nutrients, Aquatic acidification, Terrestrial ecotoxicity, etc.

Table 2 Continued ......

CML 2 baseline 2000	The CML 2 baseline method elaborates the problem-oriented (midpoint) approach. The CML Guide provides a list of impact assessment categories grouped into: 1. obligatory impact categories; 2. additional impact categories; 3. other impact categories	Depletion of abiotic resources, Climate change, human toxicity, fresh-water aquatic eco-toxicity, eutrophication, acidification, etc.

3. RESULTS AND DISCUSSION

3.1 Cost Analysis on SF-EDR and SF-UF-RO

The pilot testing provides the results including permeate flux of individual units, requirement of chemicals and consumables, and electricity consumption. More information was given in the studies of Hsu et al. [16,17]. By referring to the inquiry from the membrane module manufacturers, cost analysis was conducted to estimate the capital requirement of building large-scale plant for reclaiming the effluent. From the preliminary plans, the supply of reclaimed effluent from Futian to the industrial parks in neighborhood is 30,000 CMD. Thus the subsequent evaluation is based on this production scale. For a 30,000 CMD plant, our estimation shows that it costs US$ 0.65 to produce one cubic meter of filtrate from the SF-UF-RO process, including building cost of US$ 0.35 (land construction + mechanical and electrical equipments, depreciation period 20 years) and the operation/maintenance cost of US$ 0.30. On the other hand, SF-EDR costs US$ 0.57 to produce one cubic meter of filtrate, including building cost of US$ 0.33 and the operation/maintenance cost of US$ 0.24. From this angle, wastewater reclamation using SF-EDR costs less than using SF-UF-RO, though SF-UF-RO produces filtrate of better quality. The main reason that SF-EDR is cheaper than SF-UF-RO is that EDR does not require an expensive pretreatment like UF. On the other hand, EDR requires more frequent cleaning (once every month) than that of RO (once every three to four months). The high cost in chemicals and staffs for maintaining EDR offsets the advantage that EDR does not need UF as pretreatment.

3.2 Environmental Impacts Assessment on SF-EDR and SF-UF-RO

We used the four methods (or the "index system"), Eco-indicator 99, Ecopoints 97, Impact 2002+, and CML 2 baseline 2000 to quantify the aforementioned environmental impacts of two reclamation processes. Using different index systems usually gives inconsistent rankings. It also provides more comprehensive view for evaluating the potential impacts of the two processes [8].

Using Eco-indicator 99 gave the results in Fig. 3 (percentage characterization chart). The process "SF-UF-RO" had fewer impacts on the counted impact categories than "SF-EDR" to produce one ton of pure water, especially regarding climate change, adsorbed inorganics, and acidification/ eutrophication. The percentage characterization charts of Ecopoints 97 and Impact 2002+ are illustrated in Fig. 4 and Fig. 5, respectively. Both methods show similar results to that of Eco-indicator 99. Process "SF-UF-RO" had fewer impacts in most categories than those of "SF-EDR", except the categories of eutrophication substances (ammonia NH_3 and phosphate P) of Ecopoints 97 and non-carcinogens of Impact 2002+.

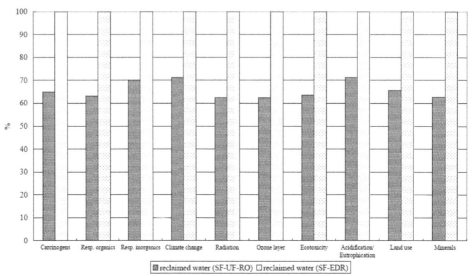

Comparing 1 kg 'reclaimed water (SF-UF-RO)' with 1 kg 'reclaimed water (SF-EDR)'; Method: Eco-indicator 99 (I) V2.04 / Europe EI 99 I/A / characterization

Fig. 3a. Percentage characterization chart of method eco-indicator 99

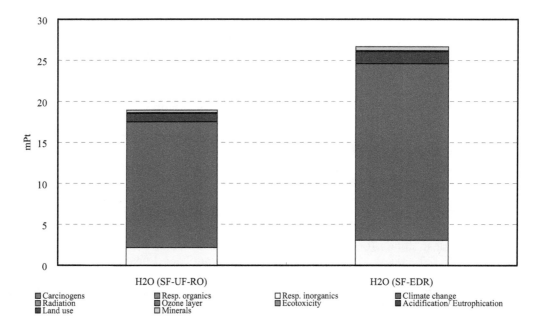

Fig. 3b. Overall scores of method eco-indicator 99

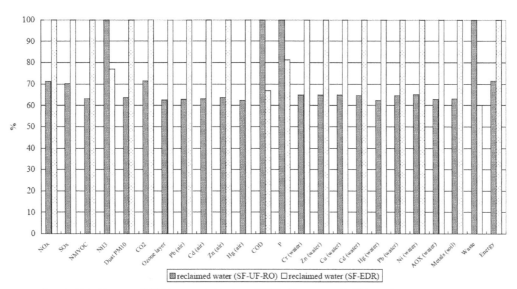

Comparing 1 kg 'reclaimed water (SF-UF-RO)' with 1 kg 'reclaimed water (SF-EDR)'; Method: copoints 97 (CH) V2.06 / Ecopoints / characterization

Fig. 4a. Percentage characterization chart of method ecopoints 97

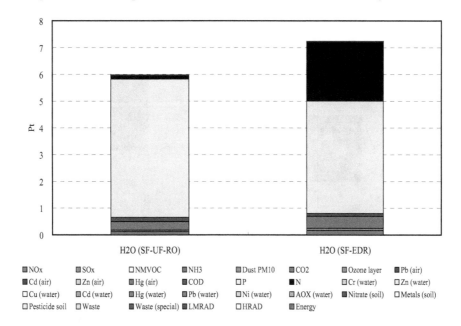

Fig. 4b. Overall Scores of Method Ecopoints 97

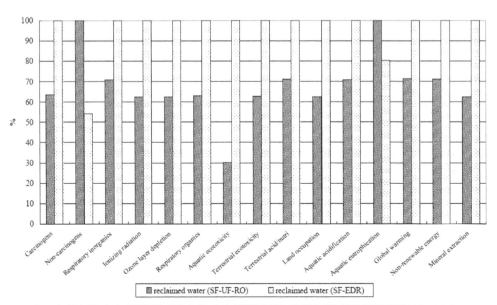

Comparing 1 kg 'reclaimed water (SF-UF-RO)' with 1 kg 'reclaimed water (SF-EDR)'; Method: IMPACT 2002+ V2.05 / IMPACT 2002+ / characterization

Fig. 5a. Percentage characterization chart of method impact 2002+

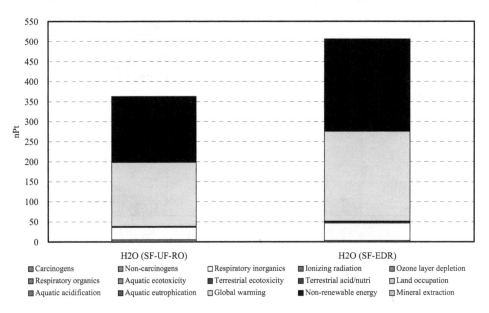

Fig. 5b. Overall scores of method impact 2002+

On the other hand, CML 2 baseline 2000 gave different results from the other three methods (Fig. 6). Process "SF-EDR" impacts less in the categories of human toxicity, fresh-water aquatic eco-toxicity and marine aquatic ecotoxicity than those of "SF-UF-RO". It is related to more concentrated pollutants in the retentate from RO due to its higher rejection ratio to the substances in the influent.

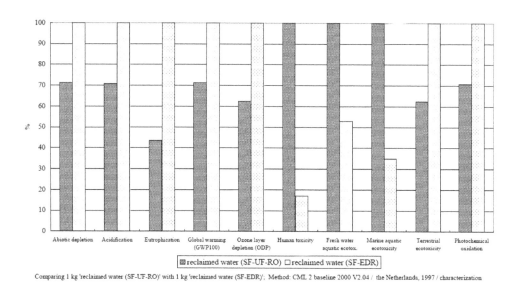

Comparing 1 kg 'reclaimed water (SF-UF-RO)' with 1 kg 'reclaimed water (SF-EDR)'; Method: CML 2 baseline 2000 V2.04 / the Netherlands, 1997 / characterization

Fig. 6a. Percentage Characterization Chart of Method Impact CML 2 Baseline 2000

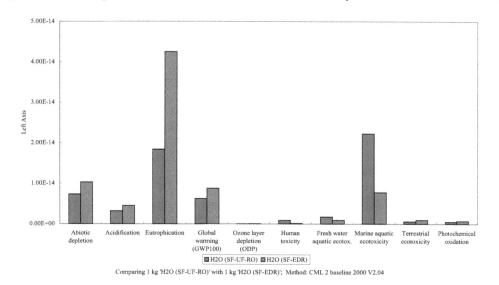

Comparing 1 kg 'H2O (SF-UF-RO)' with 1 kg 'H2O (SF-EDR)'; Method: CML 2 baseline 2000 V2.04

Fig. 6b. Normalization chart of method impact CML 2 baseline 2000

Accordingly, three of the four methods demonstrate that SF-EDR has more significant environmental impacts than SF-UF-RO (Table 3). The main reason is that the SF-EDR process consumes more electricity, and released more greenhouse gas than SF-UF-RO. The scores in greenhouse effects or climate change of SF-EDR are higher. On the other hands, the pretreatment SF provides less sufficient protection to EDR, and EDR requires chemical cleaning more frequently. It consumes more chemicals than SF-UF-RO and generally leads to higher scores in other fields (like respiratory inorganics, nitrogen emission and so on). Only the method "Impact CML2 baseline 2000" shows SF-UF-RO impacting more than SF-EDR due to the significant toxicity of the RO retentate. This may not be significant in the case of municipal wastewater reclamation as the composition is much simpler than the industrial wastewater. The ecological risk from toxicity issues of RO retentate may be controllable. It is concluded that the pressure-driven SF-UF-RO may be a more environmentally friendly process of effluent reclamation than SF-EDR, especially in the aspect of climate change mitigation.

Table 3. The Overall Score of Three Methods

Methods	Unit	The Overall Score of SF-UF-RO	The Overall Score of SF-EDR
Eco-indicator 99	Pt	2.505E-05	3.528E-05
Ecopoints 97	Pt	5.978E+00	7.239E+00
Impact 2002+	Pt	3.629E-07	5.064E-07

* Comparing 1 kg "reclaimed water (SF-UF-RO)" with 1 kg "reclaimed water (SF-EDR)"
** Method Impact CML2 baseline 2000 do not apply the single point score method to give an overall score.

4. CONCLUSION

In the literature, the selection of the effluent reclamation process has been generally based on the production quality and the cost. To the authors' best knowledge, this study is the first report considering not only the capital cost analysis but also environmental assessment of the two reclamation processes: SF-UF-RO and SF-EDR. We have noticed that the resulting water quality from both SF-EDR and SF-UF-RO are acceptable for supplying the general-purpose industrial use in Taiwan. As EDR requires less pre-treatment than RO and sand filter product with turbidity lower than 1.5 NTU is acceptable for EDR inlet, SF-EDR costs less than SF-UF-RO process in the production scale of 30,000 CMD. When evaluating the aspect of environmental impact assessment, we utilized the life cycle analysis concept with the software SimaPro 7.3. Applying the methods Eco-indicator 99, Ecopoints 97 and Impact 2002+ gave the conclusion that SF-UF-RO had fewer environmental impacts than SF-EDR. Although Impact CML 2 baseline 2000 gave the opposite results, the toxicity issues are expected to be less significant in this case of municipal wastewater reclamation. Based on the concerns of climate change mitigation, the process SF-UF-RO impacts the environment less as it emits fewer greenhouse gases and uses less non-renewable energy. SF-EDR may give more impacts to the environment due to more consumption on electricity and chemicals, although it may cost less financially. The results would be a relevant reference for the process selection for the scale-up construction in Taiwan, and it offers recommendations of process selection for climate change adaption in aid of environmental security.

COMPETING INTERESTS

Authors have declared that no competing interests exist.

REFERENCES

1.   Daigger GT, Crawford GV. Enhancing water system security and sustainability by incorporating centralized and decentralized water reclamation and reuse into urban water management systems. J Environ Eng Manage. 2007;17(1):1-10.
2.   Pai TY, Chang TC, Chen HH, Ouyang CF. Using grey relation analysis to evaluate the reuse potential of municipal wastewater treatment plant effluent based on quality and quantity. J Environ Eng Manage. 2010;20(2):85-90.

3. Chu CP, Jiao SR, Lin HM, Yang CH, Chung YJ. Recycling the wastewater of the industrial park in northern Taiwan using UF-RO system: In-situ pilot testing and cost analysis. J Water Supply Res T. 2007;56(8):533-540.

4. Chu CP, Jiao SR, Hung JM, Lu CJ, Chung YJ. Reclamation of the Wastewater from the Industrial Park Using Hollow-Fiber and Spiral-Wound Membranes: 50-CMD Pilot Testing and Cost Evaluation. Environ Technol. 2009;30(9):871-877.

5. Sadrzadeh M, Razmi A, Mohammadi T. Separation of monovalent, divalent and trivalent ions from wastewater at various operating conditions using electrodialysis. Desalination. 2007;205:53-61.

6. Chao YM, Liang TM. A feasibility study of industrial wastewater recovery using electrodialysis reversal. Desalination. 2008;221:433-439.

7. Pennington DW, Potting J, Finnveden G, Lindeijer E, Jolliet O, Rydberg T, et al. Life cycle assessment Part 2: Current impact assessment practice. Environ Int. 2004;30:721-739.

8. Renou S, Thomas JS, Aoustin E, Pons MN. Influence of impact assessment methods in wastewater treatment LCA. J Cleaner Prod. 2008;16:1098-1105.

9. Ortiz M, Raluy RG, Serra L, Uche J. Life cycle assessment of water treatment technologies: wastewater and water-reuse in a small town. Desalination. 2007;204:121-131.

10. Bonton A, Bouchard C, Barbeau B, Jedrzejak S. Comparative life cycle assessment of water treatment plants. Desalination. 2012;284:42-54.

11. Barrios R, Siebel M, Helm A, Bosklopper K, Gijzen H. Environmental and financial life cycle impact assessment of drinking water production at Waternet. J Cleaner Prod. 2008;16:471-476.

12. Muñoz I, Peral J, Ayllό JA, Malato S, Passarinho P, Domènec X. Life cycle assessment of a coupled solar photocatalytic-biological process for wastewater treatment. Water Research. 2006;19:3533-3540.

13. Nijdam D, Blom J, Boere JA. Environmental Life Cycle Assessment (LCA) of two advanced wastewater treatment techniques. Stud Surf Sci Catal. 1999;120(B):763-775.

14. Chatzisymeon E, Foteinis S, Mantzavinos D, Tsoutsos T. Life cycle assessment of advanced oxidation processes for olive mill wastewater treatment. J Cleaner Prod. 2013;54:229-234.

15. Oren Y, Korngold E, Daltrophe N, Messalem R, Volkman Y, Aronov L, et al. Pilot studies on high recovery BWRO-EDR for near zero liquid discharge approach. Desalination. 2010;261:321-330.

16. Hsu YC, Huang HH, Huang YD, Chu CP, Chung YJ, Huang YT. Survey on production quality of Electrodialysis reversal and reverse osmosis on municipal wastewater desalination. Water Science & Technology. 2010;66(10):2185-2193.

17. Hsu YC, Wang YH, Wu SC, Wu CM, Chu CP, Chung YJ. Adjusting chlorine dosage and to prevent bio-growth and minimize trihalomethanes in reverse osmosis filtrate in a wastewater reclamation process. CLEAN – Soil, Air, Water. 2012;40:254-261.

4

Synergistic Use of Remote Sensing for Snow Cover and Snow Water Equivalent Estimation

Jonathan Muñoz[1*], Jose Infante[1], Tarendra Lakhankar[1],
Reza Khanbilvardi[1], Peter Romanov[1], Nir Krakauer[1] and Al Powell[2]

[1]NOAA-Cooperative Remote Sensing Science and Technology Center (NOAA-CREST),
City College of New York, 160 Convent Ave, NY 10031, USA.
[2]NOAA/NESDIS/Center for Satellite Applications and Research (STAR)
5200 Auth Road, WWB, Camp Springs, MD 20746, USA.

Authors' contributions

This work was carried out in collaboration between all authors. All authors read and approved the final manuscript.

ABSTRACT

An increasing number of satellite sensors operating in the optical and microwave spectral bands represent an opportunity for utilizing multi-sensor fusion and data assimilation techniques for improving the estimation of snowpack properties using remote sensing. In this paper, the strength of a synergistic approach of leveraging optical, active and passive microwave remote sensing measurements to estimate snowpack characteristics is discussed and examples from recent work are given. Observations with each type of sensor have specific technical constraints and limitations. Optical sensor data has high spatial resolution but is limited to cloud free days, whereas passive microwave sensors have coarse spatial resolution and are sensitive to multiple snowpack properties. Multi-source and multi-temporal remote sensing data therefore hold great promise for moving the monitoring and analysis of snow toward estimates of a suite of snow properties at high spatial and temporal resolution.

Keywords: Snow; optical; active; passive; microwave; remote Sensing.

---

*Corresponding author: Email: tlakhankar@ccny.cuny.edu;

1. INTRODUCTION

Snow is a key component of Earth's energy balance, climate, environment, and a major source of freshwater in many regions. Seasonal snow typically covers 30% of the total land area of the Northern hemisphere. In the Southern Hemisphere, outside of Antarctica is generally limited to smaller regions in elevated areas, such as the Andes and the mountains of New Zealand. Snow is among the most important earth's surface characteristics and it has a complex interaction with the landscape and different atmospheric conditions. For this reason, monitoring the spatial and temporal variability of snow cover at high resolution provides valuable information for various weather and climate applications [1,2]. Consequently, accurate information about snow characteristics is required to improve the accuracy of existing hydrological and numerical weather models.

The conventional source of acquiring information on snow characteristics are reports from a network of ground-based meteorological stations at which daily observations are performed. However, most of Earth's snow is located in remote and inaccessible areas, where populations are sparse or nonexistent and extreme conditions limit the ability to monitor the snow conditions continuously. As an alternative, satellite remote sensing has been used for mapping snow cover and to estimate snow characteristics for several decades[3–5]. Satellite sensors can provide higher spatial and temporal resolutions than conventional methods. However, the research questions that still are unresolved are: What level of accuracy can be reached using satellite remote sensing, and what direction should be taken to improve existing products?

Despite the success of existing operational snow detection algorithms in mapping snow cover under most conditions, there are limitations in extending these to monitoring snow properties of interest, such as snow depth and snow water equivalent[3,6]. The existing limitations suggest that more attention should be paid on making better use of multi-source and multi-temporal remote sensing data sets. Given that different technical constrains affects each particular sensor, the synergy of satellite observations in the visible and in the microwave spectral bands is an important approach to improve the mapping and monitoring of the snow cover and snowpack properties (Fig.1.). The merging of multiple sensor observations with different spectral bands will lessen the instrumental limitations (e.g. spatial resolution, temporal resolution and all weather capabilities).Moreover the combination of different Spectral Bands adds further independent information about the snow characteristics.

This article reviews various remote sensing techniques for snow studies, such as: optical/infrared, active and passive microwave. However, special attention is given to their synergistic use that can substantially improve snow retrievals.

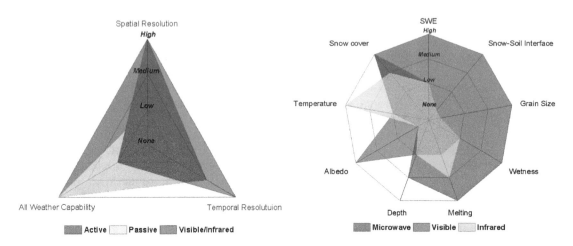

Fig.1. (a) Sensor capabilities in qualitative terms for spatial, temporal resolution and data production and (b) Sensor responses to snowpack properties. Different regions of the electromagnetic spectrum provide useful information about the snow characteristics. Nevertheless, certain regions had better capabilities or responses to measure certain properties. For this reason, the integration of all the available resources could lead to unbiased and better estimations of the snowpack properties

2. REMOTE SENSING OF SNOW

2.1 Optical Remote Sensing

Unique characteristics of snow, like its high reflectance in the visible part of the spectrum and its low reflectance in the mid-infrared is the base for most of the well-known existing techniques for snow detection using the visible (VIS) and infrared (IR) bands [7]. Furthermore, the snow reflectance on these spectral ranges differs from the reflectance of water clouds and soil making possible the foundation for automated and semi-automated snow detection techniques (Fig.2.). Furthermore snow extent is in most cases relatively straightforward to observe using visible imagery because of the high snow albedo (up to 80% or more in the visible part of the electromagnetic spectrum) relative to most land surfaces [8,9]. Overall, visible imagery is better at detecting snow cover extent than at quantifying snow characteristics like snow depth or snow water equivalent. For this reason, the primary use of optical sensors is to provide accurate and spatially detailed information on the snow cover distribution. But, on the negative side, snow mapping with the sensors using visible spectrum is impossible at night and in cloudy conditions. As a result, snow maps generated from the visible imagery can have gaps in the area coverage.

Because surface snow properties and snow cover are rapidly changing phenomena in many regions, there is a need for frequent data. Current optical sensors operated at various spatial and temporal resolutions (Table 1.), providing vital information for monitoring snow. Nevertheless user needs can be compromised, because these methods are limited by a number of factors, such as clouds, forest cover fractions, terrain heterogeneity and different atmospherics phenomena's.

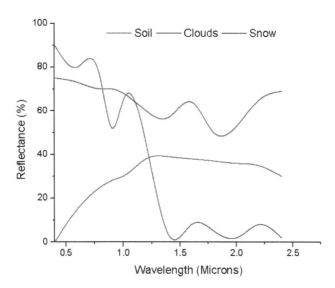

Fig.2. Representative spectral reflectance of snow, vegetation and clouds in the VIS and IR region

Table 1. Typical Sensors used for snow cover mapping

Platform sensor	Spatial resolution	Array dimensions	Temporal resolution
Aircraft (Ortho-photo)	0.5m to 3m	As needed	As needed
Landsat *TM*	28.5 Km	2.5/5	16 days
NOAA-15 (*AVHRR*)	1.1 km	200/500	12 hours
GOES (VISSR)	1.1 km	200/500	As needed
Terra & Aqua (*MODIS*)	500 m	1200	24 hours

Images from satellites have been used for mapping snow cover for several decades; for example, the USA National Oceanic and Atmospheric Administration (NOAA) began mapping snow using satellite-borne instruments in 1966. Satellite sensors such as the Advanced Very High Resolution Radiometer (AVHRR) onboard NOAA and METOP satellites, Moderate Resolution Imaging Spectro-radiometer (MODIS) onboard Earth Observing System (EOS) satellites, Imager instruments onboard Geostationary Operational Environmental Satellites (GOES) and a number of other sensors onboard polar orbiting and geostationary satellites have been used for snow cover mapping. However, most of the optical remote sensing products classify each land pixel into one of three categories, "snow", "no snow" and "undetermined", where the latter category includes cloudy pixels and pixels corresponding to night time observations. In the case of NOAA satellites their interactive maps are binary and include pixels of two classes, "snow" and "no snow" (
Fig.**3**.).

Early snow cover product using optical remote sensing used threshold-based criteria tests, decision rules and differences between radiances. For example, based on that snow is highly reflective in the visible part of the EM spectrum and highly absorptive in the near-infrared, [10] used the ratio of radiances between 1.6 - 0.754 µm channels and IR band to discriminate between snow and clouds [10]. One decade later, [11] used a more sophisticated normalized difference band ratio (NDSI) of spectral reflectance's measured with the Landsat Thematic Mapper (equation 1), Where ρ_2 and ρ_5 is in the range of 0.53- .61 μm and 1.57-1.78 μm respectively. This index is the base of several snow indexes developed in recent years.

$$NDSI = \frac{\rho_2 - \rho_5}{\rho_2 + \rho_5}$$ IfNDSI>4, classified as snow   .... (1)

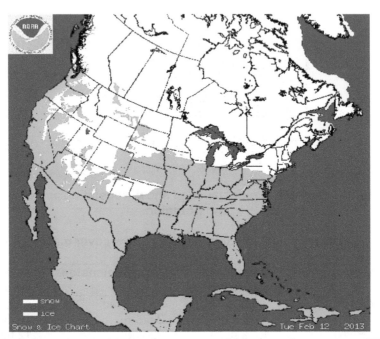

Fig.3. NOAA's interactive multi-sensor snow and ice mapping system (IMS) product shows ice and snow cover extent in the North Hemisphere

Current optical algorithms for the MODIS snow-cover products were improved and enhanced from previous operational products, allowing high resolution, daily availability and the capability to better separate snow and clouds [12]. However, snow mapping using optical wavelengths still requires clear sky conditions and sufficient daylight. For this reason and in order to mitigate these disadvantages, some authors like [7] have suggested the importance of the synergy of satellite optical (visible/infrared) and microwave data to map snow extent (**Fig. 4**.) and monitor its evolution in time and space. Some of the most well know optical snow cover products are summarized in Table 2.

Additionally to the snow cover mapping, other applications for snow studies using optical remote sensors include: derive the snow fraction within the satellite field of view and identifying snow melt using the infrared bands. The importance of fractional snow cover identification is based on the fact that the snow covered area frequently varies at a finer spatial resolution, than the one provided by the remote sensing instrument. This discrepancy, on spatial distribution cause a "mixed pixel" problem, caused by spatial mixture of snow with vegetation, soil and rocky surfaces [13]. Moreover, binary classification can failed in areas that have patchy snow such as near the snow line [14]. Essentially, map snow-covered areas at sub-pixel resolution can provide a better representation of snow distribution for the modeling community.

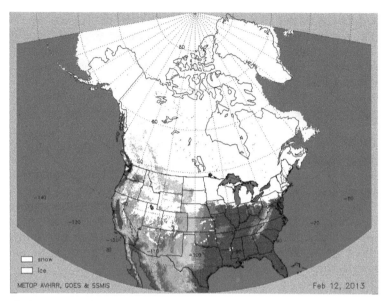

Fig. 4. Automated snow cover product for northern hemisphere [7].

Table 2. Snow covers products with their respective accuracies.

Snow cover product	Spatial resolution	Temporal resolution	Primary uses	Reference	Accuracy
IMSSystem	4km	Daily	Input to operational atmospheric forecast models.	[5]	≈ 90% to Stations
Automated snow mapping system*	4-km	Daily	Input to numerical weather models, Support IMS.	[3]	≈ 85- 90% to IMS
MODIS	500m	Daily, 8 Days & monthly	Modeling at hemispherical and regional scale.	[12]	>80% to IMS

Synergistic product from optical and microwave observations

2.2 Microwave Remote Sensing

Satellite observations in the microwave spectral range have also been used for the global monitoring of snow cover and surface snow properties for more than three decades. For example, the remote sensing community uses several data sets from the following sensors: the Electrically Scanning Microwave Radiometer (ESMR) (1973–1976), Scanning Multichannel Microwave Radiometer (SMMR) (1978–1987), Special Sensor Microwave/Imager (SSMI/S) (1978–Present) and Advanced Microwave Scanning Radiometer–Earth Observing System (AMSR-E) (2002–2011).

Also, it has been demonstrated that active microwave have similar potential as passive microwave for snowpack properties studies [15–17]. Advantages like a finer spatial resolution and more frequent data holds promise. However, the complexity of the data and the effect of surface characteristics like soils complicate its applicability at global scale. Active microwave remote sensing instruments like RADARSAT-1, QuickSCAT, CryoSAT and CryoSAT-2 are currently operational. In general, both active and passive techniques have brought new products with some advantages over counterpart's instruments in other regions of the EM spectrum.

2.2.1 Passive microwave remote sensing

Passive microwave sensors detect the weak microwave radiation that is constantly emitted from the surface and atmosphere of the Earth. In the field of microwave radiometry, the microwave radiance is mostly expressed in terms of brightness temperature, Tb at the measured frequency (Fig.5.). In the case of the snow the upwelling microwave radiation is emitted by the sub-snow surface and altered by the snowpack and consequently it carries information on the physical properties of the snowpack. Furthermore, the radiation emitted by the snowpack strongly depends on the physical properties of the snowpack, including liquid water content, snow density, grain size, vertical temperature profile and often, on the state of the ground surface beneath the snowpack [18–20].

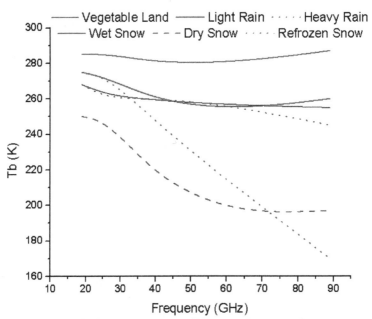

Fig.5. Microwave spectral responses at different frequencies to snow and land parameters

Currently, there are many satellite sensors measuring brightness temperature in different microwave bands. These multi-frequency observations can be used to classify snow conditions, to estimate the water equivalent of dry snow and to determine the start of the melting period. With the results of several years of ground-based microwave observations along with the experience gained with satellite-borne microwave radiometers, several microwave emission models have been developed. Many of these physically based models have been proposed to describe the relationships between the microwave emission and snowpack parameters such as the mean snow grain size, density and depth [19,21–23] (Table **3**.). These microwave emission models may be based on physical principles as well as observations and they can be classified as empirical, semi-empirical or theoretical. The objective of microwave models that are capable of predicting the measured radiation, facilitate the use of inversion techniques to estimate snow parameters. However, proper characterization of the behavior of snow-emitted microwave radiation throughout winter season remains a challenge and it is still a subject of study. Additionally, numerous research studies have used emission models along with brightness temperature at 19, 37 and 85/89 GHz microwave frequencies from satellite-mounted instruments, including AMSR-E and SSM/I (Fig. 6.), for estimation of the snow cover extent, snow depth and snow water equivalent [4,7,21,24,25].

Table 3. Characteristics of commonly used microwave emission models for snowpack property retrieval

Model	Model type	Characteristics	References
Grody	Empirical	Decision tree algorithm for global snow covers mapping from spectral gradients in SSM/I data.	Grody & Basist [4]
HUT	Semi-Empirical	Considers homogeneous snow or multiple layers. Includes the atmosphere, soil and vegetation.	Pulliainen et al.[22]
MEMLS	Semi-Empirical	Considers a layered structure of the snowpack. Classical RT with Empirical scattering and absorption properties.	Wiesmann & Mätzler [23]
DMRT	Theoretical	Based on scattering theory. Considers snowpack as a medium consisting of scattering particles.	Tsang et al.[30] & Tsang & Kong [31]

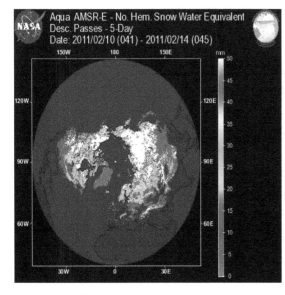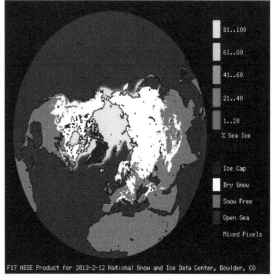

Fig. 6. (Left) AMSR-E/Aqua 5-day products of global snow water equivalent on EASE-Grids format for northern hemisphere. (Right) The Near-Real-Time SSM/I-SSMIS EASE-Grid Daily Global Ice Concentration and Snow Extent product for northern hemisphere (Source: http://nsidc.org)

The spectral gradient(difference between microwave bands) is used in most algorithms for snow cover detection [26]. Kunzi et al. [27] Used the spectral gradient method on SMMR data to map snow cover extent and found that the resulting line delimiting the snow cover corresponded to 5 cm snow depth. In addition, earlier snow depth retrieval algorithms [28,29] provided an "instantaneous" daily snow depth estimate based on differences in brightness temperature between microwave frequencies [28,29]. The same multi-frequency approach has been used for many years to retrieve SWE (Table 4.). Currently, AMSR-E (Fig. 6.) global SWE product(http://www.ghcc.msfc.nasa.gov/AMSR/) has been most frequently used by researchers.

Table 4. Basic algorithms to retrieve snow properties using microwave remote sensing

Snow depth	SWE	Snow cover
$SD = A[\Delta(Tb)]$	$SWE = A + B \frac{|\Delta(Tb)|}{f2-f1}$	$[\Delta(Tb)] > 3.8$ K
Where: $\Delta(Tb)$ is the difference in brightness temperature between 19H GHz and 37H GHz channels, (A= 1.59 cm)	Where: A and B are the offset and the slope of the regression of $f_1 = 37$H GHz and f_2=19H GHz	Where: 3.7 is the threshold for snow cover detection.

In terms of the accuracy of existing algorithms,[32] compared several passive microwave snow products including [4,26,33]. They reported that all of them underestimate snow in comparison with the NOAA Northern Hemisphere snow charts derived from manual interpretation of visible satellite data. Generally, they concluded that the microwave data indicates less snow-covered areas than the visible data throughout the year. The mean difference during the winter months (November-April) was about 4 million square kilometers, decreasing from 8 million square kilometers in November to about 0.3 million square kilometers (approximate 1% of the snow covered areas) in April [32]. The underestimation and the large difference in snow extent in early winter can be explained due to the thin snow cover, the inability of present microwave products to detect shallow (< 5cm) and discontinuous snow cover. In contrast, they explain that there is no reason to suppose that the optical data would overestimate.

Following a similar path [8], compared different snow products, including IMS and MODIS (visible), AMSR (Microwave) and the Canadian Meteorological Center Snow Product (CMC) which is a hybrid model/observational dataset, on a global basis. They conclude that there is a significant concordance among products during clear skies and non-melting conditions but large discrepancies in the presence of wet, thin and sporadic snow. Similar conclusions were drawn by [32,34]. These results suggest that the synergy of optical and microwave data can minimize the inconsistency between data sets. Visible imagery can help diagnose several of the sources of error in estimating snow depth and SWE using microwave remote sensing, including heterogeneity and variability in snow cover, grain size, snow density and snow liquid water content within the microwave sensor footprint through land cover change, patchy snow fall and/or melting snow.

2.2.2 Active microwave remote sensing

Active Microwave remote sensing, known as radar or Synthetic Aperture Radar (SAR) is based on actively transmitting a powerful pulse of microwave radiation and measuring its backscatter from the target surface. Several authors have proven that, in theory, active microwave remote sensing has similar sensitivity to snow properties as passive microwave remote sensing [35,36] and enables more precise retrievals because the output power is known. Moreover, active retrievals typically achieve much better spatial resolution than passive sensors due to their higher signal-to-noise ratio, even though this also depends on the antenna size. Active microwave measurements are very promising for snow remote sensing, but retrievals are complicated because the backscattered energy is influenced by the soil type and soil moisture as well as the geometry of the microwave beam and receiver. Moreover, SAR is very sensitive to snow melting or wet snow, providing the ability to adequately distinguish or discriminate between bare soil and wet or melting snow [37].

Instruments such as the QuikSCAT active microwave scatterometer has been used to estimate the timing of snow melt across Greenland [38] and Arctic lands [39] with fairly accurate results. Both studies were based in the backscattering's signature difference (Fig. 7) between the dry snow and wet snow. Furthermore, a product developed by [40] for mapping wet snow in mountainous terrain showed very good correlation with existing snow cover retrievals. This product used comparisons between images from consecutive passes of the Synthetic Aperture Radar (SAR). Then, filtering was performed over the measured backscattering using a high precision Digital Elevation Model (DEM) and a reference image.

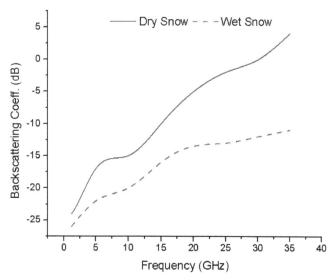

Fig. 7. Response to active microwave sensors by dry and wet snowpack conditions at different frequencies

Additionally, different methods based on active microwave sensors have been used for retrieving snow properties. Such properties are snow melting, snow depth and snow cover, which have been tested in different situations and regions around the world [21,41,42]. In some cases these products provided accurate results; however in other cases they offered poor to fair correlation when compared to in-situ measurements. For instance, [42] study compared the melt duration derived from passive microwave SSM/I's Brightness Temperature (TB) and QuikSCAT's backscattering signal time series, by making use of threshold-based processing methods. The estimation of snow depth were compared using two approaches, first a simple static approach based on radiative transfer models and second, a dynamic approach which assumes that snowpack properties are temporally and spatially variable [21]. The most important outcome from this study is that the static approach underestimated the snow depth more than the dynamic algorithms. As well, in active microwave remote sensing there are other possibilities that are being explored. For instance, P-band has been used for sea-ice detection and the X-band that has been used experimentally for high resolution mapping.

In general, it can be concluded that for the proper implementation of active microwave in snow studies, additional factors need to be considered. One example is, taking in account that the snowpack is a combination of several parameters that are continuously changing in time and space. Finally, research on active microwave remote sensing of snow illustrates a peculiar situation. It has been studied for more than three decades, including multiple field studies, theoretical models and additionally large data sets are available. However, active microwave retrievals still are not widely used on operational products.

3. SYNERGISTIC APPROACH

While standalone approaches for snow estimation using individual satellite instruments have made significant progress in recent years, many of the products currently used are still based on empirical or semi-empirical relationships and are accurate only over a limited range of snow properties [6]. For this reason, some studies have explored the possibility of improving snow retrievals by incorporating the use of multi-source and multi-temporal remote sensing data. Given the technical constraints and limitations previously discussed, the synergy of satellite observations in the visible and in the microwave spectral bands is an important approach to improve the mapping and monitoring of the snow cover and snow pack properties.

3.1 Snow Cover

Snow has a high reflectivity ratio in the optical range of the electromagnetic spectrum; this characteristic makes its detection very easy using different combinations of visible and infrared wavelengths. At the same time, visible and infrared bands have the well-known limitation of allowing only clear day detection, as neither of them can penetrate clouds. Also snowpack properties such as the snow water content can't be derived from VIS/IR wavelengths. A solution to the cloud problem and to inferring properties beyond snow cover is incorporating microwave data either passive or active that can be acquired during night or day. Although the cloud problem is removed, interpretation of microwave imagery is much more difficult compared to optical-based indices [1,43].An automatic system of snow mapping with a spatial resolution of 5 km using GOES visible and infrared data and SSM/I microwave data was developed[7,44]. This method was shown to be as precise as the IMS (Interactive Multi sensor Snow and Ice Mapping System) products if not better especially on the level of the consistency of the time series. In general they showed the utility of the multi-sensor techniques for the improvement of the snow detection. Others, have gone further, combining visible (MODIS), active (QuickSCAT) and passive (AMSR-E) microwave [45]. They demonstrate that this combination was able to distinguish between dry and wet snow using MODIS, areas where melting was at initial stages using AMSR-E and areas with heavy melting using QuickSCAT. In general these result shows that synergistic products are more accurate in determining the snow covered areas.

Future works, included: downscaling to achieve better spatial resolutions (<500m), the use of multiple instruments in different orbits to have better temporal resolution and validation. As snow cover is currently showing rapid changes because of the climate change, hydrologists and climatologist are very interested on better techniques using remote sensing for snow mapping.

3.2 Snow Water Equivalent

Passive microwave data at 19 and 37 GHz (or similar frequencies) have been historically used to retrieve snow parameters such as snow water equivalent (SWE) and snow depth. Nevertheless, the numbers of parameters that influence brightness temperatures in the microwave range make it very difficult to accurate estimate SWE. Validation studies shows that the accuracy of the AMSR-E SWE products is around 68.5% [46] and they tend to overestimate SWE [8].

There are two different approaches that can be used to increase the accuracy of existing SWE products. First approach, merging visible and microwave data with ground observations. The second approach is to explicitly combine active and passive microwave remote sensing data. Most of the studies using active microwave data have concentrated on the separation between wet and dry snow. However, the sensitivity of active microwave also exists to other parameters such as snow grain size and SWE [18].

The first approach is used by [46] to create a new SWE product with better spatial resolution (5km) and less overestimation using combination of MODIS data with the AMSR-E SWE product [46]. In their analysis of the product they conclude that the synergy produced slightly better SWE accuracy with respect to ground measurements. However, AMSR-E is no longer operational, but the same principle can be used to merge other optical data sets like GOES-R with SSM/I, AMSU or the next generation in the family the Advanced Technology Microwave Sounder (ATMS).

The second approach proposed by [36], combined active (QuikSCAT/Sea Winds) and passive (SSMI) data for monitoring key snow parameters in Finland [37]. Data from 21 test sites were used to validate their study. In general the results show that combined active and passive microwave sensors provide useful information on diurnal and seasonal variability and increase the accuracy of SWE by approximate 6%. Additionally, Azar et al. [47] demonstrate that combining SSM/I with QuikSCAT and NDVI produce improved and more accurate SWE estimates than those obtained by only using SSM/I [47]. Furthermore, both active and passive microwave are very sensitive to snow wetness. Given the complexity of the relationship between this parameter and the microwave emission is unrealistic the accurate estimation of SWE, based on the existing microwaves models. Consequentially the assimilation of surface temperature using visible and infrared bands can be used to estimate the snow wetness, resulting in better SWE estimates. Overall, there is evidence that the combination of visible, active and passive microwave data to retrieve SWE may improve the results from those obtained just using passive microwave.

4. SUMMARY

Unbiased estimates of snow properties could lead to much better understanding of hydrological processes in snow covered watersheds. This understanding is important not only for seasonal-scale processes such as the snowmelt contribution to water supply systems and snow-atmosphere interactions, but for predicting flash floods caused by rapid snowmelt. Specific future improvements and challenges for using synergistic approaches include refinement of snow-cover extent estimates with better spatial-temporal resolutions, minimizing SWE retrieval errors, and improving our ability to ingest remote sensing data into land-surface models.

To conclude, several snow characteristics can be measured using different remote sensing techniques based on different parts of the electromagnetic spectrum. The accuracy and/or sensitivity of retrievals will depend on the parameter being measured and instrument used for the snowpack estimation. Future efforts towards improving snow retrievals should therefore seek to merge various of the existing products into one capable of compensating the disadvantages of any one retrieval method, and to better understand the underlying physics of the snowpack in order to make more robust operational products, compared to current ones that are based on empirical relationships between sensor data and snow extent and properties.

ACKNOWLEDGEMENTS

This study was supported by National Oceanic and Atmospheric Administration (NOAA) under grant numbers NA06OAR4810162, NA11SEC4810004 and NA09NES4400006. The views, opinions and findings contained in this report are those of the authors and should not be construed as an official National Oceanic and Atmospheric Administration or US Government position, policy, or decision.

COMPETING INTERESTS

Authors have declared that no competing interests exist.

REFERENCES

1.   Robinson DA, Dewey KF, Heim RR. Global Snow Cover Monitoring: An Update. Bulletin of the American Meteorological Society. 1993;74(9):1689–1696.
2.   Dyer J. Snow depth and streamflow relationships in large North American watersheds. Journal of Geophysical Research. 2008;113(D18):1–12.
3.   Romanov P. Satellite-derived snow cover maps for north America: Accuracy assessment. Advances in Space Research. 2002;30(11):2455–2460.
4.   Grody NC, Basist AN. Global identification of snowcover using SSM/I measurements. IEEE Transactions on Geo-science and Remote Sensing. 1996;34(1):237–249.
5.   Helfrich SR, McNamara D, Ramsay BH, et al.. Enhancements to, and forthcoming developments in the Interactive Multisensor Snow and Ice Mapping System (IMS). Hydrological Processes. 2007;1586(12):1576–1586.
6.   Tedesco M, Derksen C, Pulliainen J. Hemispheric snow water equivalent: The need for a synergistic approach. EOS, Transactions American Geophysical Union. 2012;93(31):305.
7.   Romanov P, Gutman G, Csiszar I. Automated monitoring of snow cover over North America with multispectral satellite data. Journal of Applied Meteorology. 2000;39:1866–1880.
8.   Frei A, Tedesco M, Lee S, et al.. A review of global satellite-derived snow products. Advances in Space Research. 2012;50(8):1007–1029.

9.   Qu X, Hall A. What Controls the Strength of Snow-Albedo Feedback? Journal of Climate. 2007;20(15):3971–3981.

10. Kyle HL, Curran RJ, Barnes WL, Escoe D. A cloud physics radiometer. Third Conference on Atmospheric Radiation American Meteorological Society. 1978;107.

11. Dozier J. Spectral signature of alpine snow cover from the Landsat Thematic Mapper. Remote Sensing of Environment. 1989;28(August 1988):9–22.

12. Hall DK, Riggs GA. Accuracy assessment of the MODIS snow products. Hydrological Processes. 2007;21(12):1534–1547.

13. Dozier J, Painter TH, Rittger K, Frew JE. Time–space continuity of daily maps of fractional snow cover and albedo from MODIS. Advances in Water Resources. 2008;31(11):1515–1526.

14. Nolin AW. Recent advances in remote sensing of seasonal snow. Journal Of Glaciology. 2010;56(200):1141–1150.

15. Stiles WH, Ulaby FT. Dielectric properties of snow. Journal of Geophysical Research. 1981;85(C2):82–18.

16. Matzler C, Aebischer H, Schanda E. Microwave dielectric properties of surface snow. IEEE Journal of Oceanic Engineering. 1984;9(5):366–371.

17. Rott H. The analysis of backscattering properties from SAR data of mountain regions. IEEE Journal of Oceanic Engineering. 1984;9(5):347–355.

18. Ulaby FT, Stiles WH. The Active and Passive Microwave Response to Snow Parameters 2. Water Equivalent of Dry Snow. Journal of Geophysical Research. 1980;85(C2):1045–1049.

19. Grody N. Relationship between snow parameters and microwave satellite measurements: Theory compared with Advanced Microwave Sounding Unit observations from 23 to 150 GHz. Journal of Geophysical Research. 2008;113(D22):1–17.

20. Lakhankar TY, Muñoz J, Romanov P, et al. CREST-Snow Field Experiment: analysis of snowpack properties using multi-frequency microwave remote sensing data. Hydrology and Earth System Sciences. 2013;17(2):783–793.

21. Kelly RE, Chang A, Tsang L, Foster JL. A prototype AMSR-E global snow area and snow depth algorithm. IEEE Transactions on Geoscience and Remote Sensing. 2003;41(2):230–242.

22. Pulliainen J, Grandell J, Hallikainen MT. HUT snow emission model and its applicability to snow water equivalent retrieval. IEEE Trans. on Geoscience and Remote Sensing. 1999;37(3):1378–1390.

23. Wiesmann A, Mätzler C. Microwave Emission Model of Layered Snowpacks. Remote Sensing of Environment. 1999;70(3):307–316.

24. Durand M, Kim EJ, Margulis SA. Quantifying Uncertainty in Modeling Snow Microwave Radiance for a Mountain Snowpack at the Point-Scale, Including Stratigraphic Effects. IEEE Transactions on Geoscience and Remote Sensing. 2008;46(6):1753–1767.

25. Simic A, Fernandes R, Brown R, et al.. Validation of VEGETATION, MODIS, and GOES+ SSM/I snow-cover products over Canada based on surface snow depth observations. Hydrological Processes. 2004;18(6):1089–1104.

26. Chang ATC, Foster JL, Hall DK. Nimbus-7 SMMR derived global snow cover parameters. Annals of Glaciology. 1987;9:39–44.

27. Kunzi KF, Patil S, Rott H. Snow-Cover Parameters Retrieved from Nimbus-7 Scanning Multichannel Microwave Radiometer (SMMR) Data. IEEE Transactions on Geoscience and Remote Sensing. 1982;GE-20(4):452–467.

28. Chang ATC, Kelly REJ, Foster JL, Hall DK. Global SWE monitoring using AMSR-E data. Geoscience and Remote Sensing Symposium, 2003. IGARSS '03. Proceedings. 2003;1(C):680–682.

29. Foster JL, Chang ATC, Hall DK. Comparison of Snow Mass Estimates from a Prototype Passive Microwave Snow Algorithm, a Revised Algorithm and a Snow Depth Climatology. Remote Sensing of Environment. 1997;62(2):132–142.

30. Tsang L, Ding KH, Wen B. Dense media radiative transfer theory for dense discrete random media with particles of multiple sizes and permittivities. Progress In Electromagnetics Research. 1992;6:181–230.

31. Tsang L, Kong JA. Scattering of Electromagnetic Waves, 3 Volume Set, Wiley-Interscience, 2001.

32. Armstrong RL, Brodzik MJ. Hemispheric-scale comparison and evaluation of passive-microwave snow algorithms. Annals Of Glaciology. 2002;34(1):38–44.

33. Goodison BE. Determination of Areal Snow Water Equivalent on the Canadian Prairies using Passive Microwave Satellite Data. 12th Canadian Symposium on Remote Sensing Geoscience and Remote Sensing Symposium. 1989;3:1243–1246.

34. Brown RD, Robinson DA. Northern Hemisphere spring snow cover variability and change over 1922–2010 including an assessment of uncertainty. The Cryosphere, 2011;5(1):219–229.

35. Matzler C, Schanda E, Good W. Towards the Definition of Optimum Sensor Specifications for Microwave Remote Sensing of Snow. IEEE Transactions on Geoscience and Remote Sensing. 1982;20(1):57–66.

36. Rott H. Synthetic aperture radar capabilities for snow and glacier monitoring. Advances in Space Research. 1984;4(11):241–246.

37. Hallikainen MT, Halme P, Takala M, Pulliainen J. Combined active and passive microwave remote sensing of snow in Finland, Ieee. 2003;2(C):830–832.

38. Nghiem S V, Tsai W-YTW-Y. Global snow cover monitoring with spaceborne Ku-band scatterometer. IEEE Transactions on Geoscience and Remote Sensing. 2001;39(10):2118–2134.

39. Wang L, Derksen C, Brown R. Detection of pan-Arctic terrestrial snowmelt from QuikSCAT, 2000–2005. Remote Sensing of Environment 2008; 112(10):3794–3805.

40. Nagler T, Rott H. Retrieval of wet snow by means of multitemporal SAR data, IEEE. 2000;38(2):754–765.

41. Yueh S, Cline D, Elder K. Airborne Ku-Band Polarimetric Radar Remote Sensing of Terrestrial Snow Cover. IEEE Transactions on Geoscience and Remote Sensing. 2008;47(10):3347–3364.

42. Dupont F, Royer A, Langlois A, et al.. Monitoring the melt season length of the Barnes Ice Cap over the 1979–2010 period using active and passive microwave remote sensing data. Hydrological Processes. 2012;26(17):2643–2652.

43. Roshani N, Zouj M, Rezaei Y, Nikfar M. Snow Mapping of Alamchal Glacier Using Remote Sensing Data. The International Archives of the Photogrammetry, Remote Sensing and Spatial Information Sciences. 2008;37(2):805–808.

44. Romanov P, Tarpley D. Enhanced algorithm for estimating snow depth from geostationary satellites. Remote Sensing of Environment. 2007;108(1):97–110.

45. Foster JL, Hall DK, Eylander JB, et al.. A blended global snow product using visible, passive microwave and scatterometer satellite data. International Journal of Remote Sensing. 2011;32(5):1371–1395.

46. Gao Y, Xie H, Lu N, et al.. Toward advanced daily cloud-free snow cover and snow water equivalent products from Terra–Aqua MODIS and Aqua AMSR-E measurements. Journal of Hydrology. 2010;385(1-4):23–35.

47.   Azar AE, Ghedira H, Lakhankar T, Khanbilvardi R. Improvement in Estimating Snowpack Properties with SSM/I Data and Land Cover Using Artificial Neural Networks, 2006. doi:10.1109/MICRAD.2006.1677080.

5

Soil Microbial Biomass Carbon, Nitrogen and Sulphur as Affected By Different Land Uses in Seronga, Okavango Delta, Botswana

T. Mubyana-John[1*] and W. R. L. Masamba[2]

[1]Department of Biological Sciences, University of Botswana, P/Bag 0022, Gaborone, Botswana.
[2]Okavango Research Institute, University of Botswana, P/Bag 285, Maun, Botswana.

Authors' contributions

This work was carried out in collaboration between both authors. Both authors read and approved the final manuscript.

ABSTRACT

Aim: The Okavango Delta at Seronga is fragmented into different land uses ranging from grasslands to woodland (Ximenia and mopane), often punctuated with cropped and fallow fields. The influence of land uses on surface (A_1 horizon) soil physico-characteristics, nitrogen, sulphur, carbon, microbial population and biomass were studied to understand soil variability in order to devise conservation strategies for the area.
Methodology: Total soil nitrogen (N) was analysed using a Leco N analyser, total carbon and sulphur by CS800 Carbon–Sulphur analyser. NH_4^+-N, NO_3^- and NO_2^- were extracted with KCl and determined using the indophenol blue method and by Griess-Ilosvay colorimetric method respectively. Microbial populations were determined by plate count method. Biomass carbon and flush of nitrogen were determined by fumigation and re-inoculation technique.
Results: All the soils had a high sand content (> 85%). Total soil N was generally very low, 0.017% in grasslands closest to the channel, 0.013% in cropped fields, 0.007% in fallow and lowest in woodlands (0.002%). Grasslands showed higher NH_4^+-N indicating low nitrification potential. Even if mopane woodlands had low total N, they had higher NH_4^+-N (0.067 ppm) and low NO_2^- compared to other land uses, this could be attributed to their inherent nitrification inhibition ability. No NO_3^--N was detected in these soils, probably due to the low nitrification ability and high leaching capacity of sandy soils.

Corresponding author: Email: mubyanat@mopipi.ub.bw;

Microbial biomass C and population were highest in the grasslands and cultivated soils, while the woodlands had lower levels.

Conclusion: Seronga soils have very low N, with the least in the woodlands furthest from floodplains. Grasslands closest to the channel basin had significantly higher total N, C and microbial biomass C but low S as opposed to the woodlands further from the channel. Cultivated areas had increased N and C levels and microbial biomass C compared to the woodland probably due to incorporation of crop residues and animal manure. The paucity of nitrifiers and undetectable NO_3^--N indicate a low nitrification potential and a high leaching ability of the soils. Fallowing of fields resulted in a decline in nutrient status.

Keywords: Mopane soils; Ximenia woodlands; grassland; soil nitrogen; microbial biomass.

1. INTRODUCTION

Seronga is the largest village on the eastern part of the Okavango Delta. The Okavango is a large, landlocked delta in the north western part of the semi-arid Kalahari basin in Botswana. It covers about 22,000 km^2, of which approximately 6000 km^2 is permanent swamp. The remaining 10,000 - 16,000 km^2 flood plain habours a high density and diversity of fauna and flora [1]. This unique system draws its water from the high rainfall highlands of Angola, as the tributaries converge into the Okavango River and then spread as an alluvial fan. The Okavango is a very important system in the country as it is the only permanent source of water in an otherwise semi-arid region. Due to its hydrologically unique nature, the Okavango Delta is bordered by many villages which depend on it for livelihood. These include fishing, crop farming, cattle rearing and recently, tourism. Tourism as a livelihood has not yet been well established in this part of The Okavango Delta. Although the rich grasslands along the channel may provide rich grazing land and a lot of water for the cattle and other domestic animals, these do not render a cash income source for the people; mostly due to distance to the market and the occasional outbreaks of foot and mouth disease of cattle. Therefore, irrespective of the sandy soils, the people rely on crop farming for their livelihood. The main crops are sorghum (*Sorghum bicolor*), maize (*Zea mays*), beans (*Phaseolus* spp), groundnuts (*Arachis hyphogea*) and water melon (*Citrullus lanatus*). These crops are often grown in mixed farming.

Although the vegetation in the Okavango floodplains consists mostly of densely populated grass species communities, which vary depending on water gradient, flood regime and grazing pressure, these grass communities grow on sandy soils (>85% sand) with low cation exchange capacity (<5meq/100g soil) [2]. The combined effect of sandy soils, low cation exchange capacity and seasonal surface flooding is likely to result in leaching of essential elements such as nitrogen, which in turn could lead to poor plant growth [3]. As such, in most cases the people tend to practice shifting cultivation to try and get to the fertile lands. The new fields cleared may either be grasslands close to the water channel bank or the woodland areas consisting of mopane (*Colosphospermum mopane*) or sour plum (*Ximenia americana*) further from the channel. Once crop yields are low, the fields are normally abandoned and left fallow for some years. The abandonment of old fields and clearing of new ones is done without knowledge of the nutrient status of the fields and technical knowhow of the necessity to conserve the ecosystem. Thus an understanding of the nutrient status of these different land use systems would help provide information necessary in planning towards the most sustainable land use.

Nitrogen (N) is an essential macro element required for plant growth and soil microorganisms. Microorganisms are also involved in many N transformations that take place in soil. These include nitrogen fixation, nitrification, mineralisation of organic residues and denitrification [4]. Plants need N in large quantities and low concentrations of it in soil leads to poor plant growth [5]. Nitrogen occurs in many forms in soil. These include ammonium-N (NH_4^+), nitrate-N (NO_3^-), nitrite-N (NO_2^-), and in organic forms where it is often part of nucleic acids, proteins, amino acids and other amino forms [3]. Thus, the role of nitrogen transforming microorganisms in Seronga soils is of great significance as they determine mineralization of vegetation to release other N forms [5]. Microbial transformations such as nitrification determine the availability of nitrogen to the plants. In some Seronga soils under mopane vegetation, due to its allelopath nature [6] plant diversity and density is extremely low, therefore very few other plants grow. An understanding of the microbial mineralisation and nitrification levels could shed light on the future cropping lands. Although denitrification may be insignificant in sandy soils due to its high aeration, this could be important in the grasslands bordering the channel. Thus, the main objective of this study was to determine soil carbon, sulphur and nitrogen forms, transformations and microbial population and biomass that occur in surface soils (A_1) in the different land use systems of Seronga.

2. MATERIALS AND METHODS

Soil samples from the four main land uses of Seronga i.e., cropped, fallow, grassland, woodland consisting of mopane and Ximenia were analysed for different soil parameters i.e., pH, texture, moisture, S, C, N and its transformations and microbial diversity and biomass.

2.1 Site and Soil Sampling

Seronga, in Botswana (S -18.816°, E 22.415°) on the north east bank of the Okavango Delta was selected as the study site due to the strong landscape contrast between the flood plains and the higher bush veld with the villages and fields stretching along the embankment. The village was chosen because it exhibits all the four land uses typical of the Okavango Delta. Surrounded by cropped fields of mostly sorghum, millet, maize, groundnuts, beans and water melons. The village is also bordered by dense woodlands consisting of mopane (*Colosphospermum mopane*), silver terminalia (*Terminalia sericea*) and sour plum (*Ximenia americana*). These woodlands harbour a lot of wild animals ranging from elephants to small animals, such as jackals and phuduhudu (*Raphicerus campestris*). Adjacent to cultivated fields are the abandoned fields that have been left fallow and are often covered with herbs such as stink weed and wild basil. Close to the channel are the grasslands. The *Cynodon dactylon* grasslands are located higher than other grasses on the banks of the channel and are used mostly for grazing.

Representative land use areas were selected and three sites with a minimum of 5 m apart in each land use were also chosen for sampling. For sampling, three sub soil samples were collected from each site using a depth marked auger to 10 cm depth and combined in one bag to make one replicate of approximately 300 g. The three replicates of at least 5m apart from the same site were labelled immediately and put in one larger bag. Three replicates were sampled from each land use. Soils from a total of 30 sampling sites were collected at 0-10 cm depth from the different land use systems i.e., cropped fields, woodland mopane, woodland Ximenia, fallow and *Cynodon dactylon* grassland. Of the different land uses, the closest to the channel are the grassland, followed by the woodland Ximenia and eventually the mopane woodland. The cultivated and fallow fields are scattered among the Ximenia and mopane woodlands. The collected samples were packed into plastic bags then cooler boxes and transported to the laboratory immediately. Once in the laboratory the portions of the samples for microbial analyses were separated from the rest and stored in the refrigerator at 4°C until use. The soil samples for nutrient analyses were air-dried, sieved through a 2 mm sieve and stored at room temperature until analyses.

2.2 Soil Physical-Chemical Characteristics

Soil moisture was determined gravimetrically and calculated from weight loss after oven drying the samples at 105°C overnight and then expressed on dry weight basis [7]. The hydrometer method of *Bouyoucos* was used to determine the sand, silt and clay content and then determining the soil textural class by using the soil textural triangle [7]. Soil pH of the samples from the different land uses was determined in water for active acidity and in 0.01M $CaCl_2$ for potential acidity. Replicate 10 g soil samples of each were mixed with 20 ml of the solution (water or $CaCl_2$). The mixture was stirred and then left to stand for 30 minutes while stirring occasionally and the pH of each soil sample was then read on a Corning scale pH meter electrode (model 215) and recorded.

2.3 Soil Carbon, Sulphur and Nitrogen and Its Components

The total soil carbon and sulphur contents of the different samples were determined using a CS 800 Carbon & Sulphur Analyser (Ultra CS800-Sci Lab UK). The CS 800 is a computer controlled carbon and sulphur analyser designed for rapid simultaneous determination of carbon and sulphur in soil and other samples. The equipment combusts the sample at 1350°C and uses Mintek reference materials containing 0.56% S and 4.0% C as standards. Total N was determined using an automatic N analyser (EA 1100, Thermo Quest). The system was calibrated using EDTA as a standard.

The NH_4^+ in the soil was first extracted with 2M KCl and then the filtered extract was analysed for NH_4^+ using the indophenol blue method described by Keeney and Nelson [8]. The intensity of the blue colour that developed from sample extracts and standards was measured calorimetrically using a spectrophotometer at 636 nm wavelength. The readings from the standards were used to prepare a calibration curve which was used for the determination of sample concentrations. A solution of 2 M KCl processed in the same way as the soil extract samples was used as the blank.

Nitrite (NO_2^-) nitrogen was determined using the modified Griess-Ilosvay method [8]. The method involves extraction of nitrite with 2 M KCl, and then addition of a diatozing reagent (sulphanilamide) in HCl to convert the NO_2^- into a diazonium salt which is later treated with a coupling reagent N-ethylenediamine to convert it to an azo compound. The red colour of the azo compound was then measured on the spectrophotometer at 540nm. The NO_2^- content in the samples was determined with reference to a standard curve that had been prepared in a similar manner but using samples containing 1-5µg NO_2^--N. The NO_3^- in the KCl extract was determined by first reducing it to NO_2^- by passing it through a copperized Cd column and the resulting NO_2^- quantified using the Griess-Ilosvay method as above [8].

2.4 Soil Microbial Diversity and Biomass

Fungal, bacterial and actinomycetes populations were determined by plate count technique on different solid agar media. Soil serial dilutions of up to 10^5 from the different land uses were made in sterile tap water and then plated on respective sterile solid agar media. For total bacteria plating was done on Trypticase soy broth (BIOMEREUX Y42830) amended with 15 g/lagar (High Media M290), fungi on potato dextrose agar and incubated at 25°C for 72 h. The actinomycetes populations were enumerated by spread plating the dilutions on starch casein agar [9] and then incubating at 25°C for 14 days, to obtain ashy-like colonies typical of actinomycetes. In all cases, after incubation colony counting was done using a colony counter and recorded for each land use. The most probable number (MPN) of biophagic protozoan was determined using the baited plate technique as outlined by Gupta and Germida [10]. Soil dilutions (10^{-2} to 10^{-5}) were plated on to the 24 multi-well MPN plates containing 0.8% NaCl solid agar (15%) and over laid with 0.5 ml of concentrated cell suspension of *Enterobacter aerogenes*as prey and incubated at 25°C and observed microscopically from 8 to 14 days.

Soil nitrification plays a significant role in converting soil NH_4^+ nitrogen to a plant available form (NO_3^-). Thus nitrification potential of these soils was estimated by determining the most probable number (MPN) of microorganisms capable of converting NH_4^+ to NO_3^- [12]. Serial dilutions (10^{-1} to 10^{-5}) of soil were inoculated into sterile 4X6 MPN plates containing inorganic NH_4^+ medium as a sole source of nitrogen for the nitrifying population capable of converting NH_4^+ to NO_3^-. While the NO_2^-oxidisers were determined by inoculating the soil dilutions into sterile inorganic medium containing NO_2^- as a sole source of nitrogen. The samples were incubated for 4 weeks at 25°C. After the incubation period the NH_4^+ medium samples were tested for the presence of NO_2^- using the Griessllosvay reagent [11] while the NO_2^- samples were tested for the presence of NO_3^- using the zinc-copper manganese dioxide powder (Zn-Cu-MnO$_2$). The assumptions made were that if *Nitrosomonas* were present in the soil dilutions inoculated into the NH_4^+ medium, it oxidizes to NO_2^- and gives a change of colour when tested with the Griessllosvay reagent. Meanwhile in the NO_2^- tubes if it has been oxidized to NO_3^- the presence of NO_3^- was detected by the Zn-Cu-MnO$_2$ [11]. The NH_4^+ and NO_2^- oxidising bacteria population were interpreted with reference to the MPN table of Cochran [12] for use with ten-fold dilutions and five tubes per dilution.

Biomass C was determined using the fumigation re-inoculation technique outlined by Jenkinson and Powlson [13] and calculated as the difference between fumigated and un-fumigated sample using a K value of 0.45 [13]. Biomass N which represents the largest portion of organic N in soils in this case was calculated as a flush of N after fumigation and re-inoculation and then determining the amount of biomass C and extrapolating to Flush of N. The flush of N was calculated from the formula: Flush of N=Biomass C/9 [14].

2.5 Statistical Analysis

Analysis of variance was performed using the SPSS 11 package. *Post hoc* analyses were performed using the Tukey Test. In the analysis, group separation was based on land use (grassland, woodland Ximenia, woodland mopane, cultivated and fallow), and the parameter studied.

3. RESULTS

3.1 Soil Physical-Chemical Characteristics in the Different Land Uses

Soil textural analysis of the different land uses showed that all the soils contained more than 85% sand and were thus classified as sandy in the soil textural triangle (Table 1) [7]. The soil pH ranged from 7.0 in the Cynodon grassland near the channel basin to 5.28 in the mopane woodland. Generally there was a decrease in soil pH as distance from the channel basin increased, with the mopane woodland soils being the most acidic. The cultivated fields generally had higher pH than the fallow fields.

Table 1. Physical land chemical properties of soil from the different land uses¥

Land use	% Soil moisture	Textural class	Soil pH	
			Active acidity	Potential acidity
Grassland	3.89±0.02b	Sandy	7.oo	6.23
Woodland Ximenia	5.77±0.01b	Sandy	5.74	5.3
Woodland mopane	2.88±0.02ab	Sandy	5.61	5.28
Cultivated	50±0.01a	Sandy	7.oo	6.22
Fallow	5.25±0.01b	Sandy	6.36	6.15

¥Values given are mean ± SD (*).
Means followed by the same letter are not significantly different at 5% level.
Moisture content on oven dry weight basis.

Soil moisture analyses indicated that these soils are generally dry as they contained moisture contents ranging from 1.50 to 5.77 % on dry weight basis (Table 1). Not surprising though because sandy soils generally exhibit low moisture holding capacity. Although the soil moisture contents were low in all the land uses, the fallow fields had significantly higher moisture content than the ploughed fields confirming the Canadian farming systems where some fields are left fallow to conserve soil moisture in some years [16].

3.2 Soil Nitrogen Components and their Microbial Transformation.

Table 2 shows the total carbon, sulphur and nitrogen content of the different land uses in the Okavango Delta soils at Seronga.

Table 2. Soil carbon, sulphur and nitrogen levels in the A_1 horizon of the different land uses¥

Land use	% C	%S	% Total N	ppm NH_4^+	ppm NO_2
Grassland	1.007±0.3c	0.002±0.1a	0.017±0.70c	0.816±0.12c	0.070±0.2c
Woodland *Ximenia*	0.617±0.1b	0.007±0.2c	0.002±1.87a	0.394±0.11a	0.025±0.1b
Woodland mopane	0.529±0.1a	0.004±0.3b	0.002±1.79a	0.671±0.12b	0.010±0.1a
Cultivated	0.719±0.2b	0.006±0.2c	0.013±0.04c	0.430±0.11a	0.037±0.2b
Fallow	0.581±0.1ab	0.004±0.1b	0.007±0.04b	0.647±0.13b	0.011±0.1a

¥Values given are means ± SD x 10^{-2}; NO_3^- content was below the 10^{-4} ppm detection level. Means followed by the same letter are not significantly different at (P=.05)

Percentage soil carbon and sulphur differed with land use (Table 2). The carbon content was significantly higher *(P=.05)* in the grassland close to the channel than in other land use systems. The cultivated fields although scattered among the woodlands also exhibited a higher C content than the fallow and the woodland systems. The high carbon content in both the grassland and cultivated land may be attributed to the high root mass in the A_1 horizons. This carbon may be contained in organic matter or root exudates due to the heavy root masses. The cultivated land's high carbon content may also be attributed to the crop residues and manure incorporation, which with time can increase soil organic matter [16]. On the contrary the woodlands due to their deep root systems showed low carbon contents in the A_1 horizon. There was a significant difference *(P=.05)* in the carbon content of cultivated and fallow fields, with lower levels occurring in the fallow. This may be occurring due microbial grazing of organic matter without replacement as in cultivated fields where there is annual amendment of cattle manure and crop residues.

The % total N in Seronga soils ranged from 0.017 to 0.002 % (Table 2). It was relatively low in all the land uses when compared to 0.02-0.05 % in other Delta soils studied elsewhere [17,18]. Lowest mean values of 0.002 % in the woodland soils and highest mean values of 0.017% in the grasslands closest to channel bankwere observed. The slightly higher N content observed in the grassland in this study could be attributed to the N fixing ability of *Cynodon dactylon* [19]. Studies from other areas indicate that this grass harbours diazotrophs such as *Azospirillum* which may fix N in the rhizosphere. When N is fixed by diazotrophs this N is usually in the NH_4^+ form. This is in agreement with this study data that shows that the NH_4^+ N was highest in the grasslands. Cultivated fields also showed slightly higher total N than the woodlands, a parameter which may be attributed to N fixation by legumes. In Seronga, most farmers do not have access to chemical fertilizers. As such, they amend their soils with cattle manure which is often rich in total N toenrich their soils [16]. The farmers also practice mixed cropping of cereals and legumes such as groundnuts (*Arachishyphogea*) and beans (*Phaseolus* spp). The N fixing ability of legumes such as groundnuts has long been understood [20]. Incorporation of crop residues often containing groundnuts, beans and water melon straw is also a common practice in these areas. Thus the combined effect of mixed cropping with legumes and manure amendments may be the factors that lead to higher total N in the cultivated fields as opposed to the uncultivated woodlands. Once the fields are abandoned the fields turn to fallow, there is neither N additions due to cultivation of legumes nor manure amendments, thus the low total N observed in the fallow fields (Table 2). Except for cultivated fields, generally total N decreased as distance from the floodplains increased. Studies by Omari et al. [21] indicate that most of the N in Okavango soils is of flood origin. The woodlands are located further from the floodplains (wetland) and receive less flood water and alluvial deposits of nutrients hence the less total N observed in these soils.

Although total N is important in soils, it is important to know the forms in which the N occurs as some N forms are more plant available than others. Furthermore plants take up N mostly in the form of nitrates. To a certain extent in acidic soils, some plants may take in N in the form of NH_4^+ [22]. Nitrogen forms such as NH_4^+ although not plant available, serve as N reservoirs for plants and microorganism. The grassland had significantly higher *(P=.05)* NH_4^+ content (0.82 ppm), followed by woodland mopane, fallow, cultivated and finally woodland *Ximenia* had the lowest content (0.43 ppm).The higher NH_4^+ observed in the grasslands may be attributed to asymbiotic N fixation potential of *Cynodon dactylon* and the low nitrification potential associated with ethylene, a nitrification inhibitor produced by grassland soils under aerobic conditions [23]. The Cynodon grasslands studied are located along the floodplains of the Okavango delta, which are often grazed by cattle. The results agreed with those of

Bonyongo and Mubyana [24] who stated that the Okavango grasslands contain more organic matter due to faecal matter from animals that graze in the grassland. They also state that the area along the floodplains receive more nutrient deposition from the surface floods of the Okavango delta. Therefore NH_4^+ is also most likely to be released from mineralisation of organic material such as roots in grassland, hence the high content of NH_4^+ recorded in the grassland. Because soil moisture plays a major role in microbial mineralisation and transformation of elements [5,11], the low moisture content in cropped fields (1.50 %) (Table 1) could also explain the low NH_4^+-N content because low soil moisture also favours the formation of insoluble nutrient-containing compounds. The Grasslands total N and NH4+-N contents could also be attributed to the low bacteria populations indicating the presence of negative parameters affecting ammonization and nitrification processes. On the contrary, this could explain the high content of nitrite ion (NO_2^-) obtained in Grasslands area.

Even if the mopane woodlands had low total N, these woodlands contained higher NH_4^+ when compared to the other land uses apart from grasslands (Table 1). In this study, this was not just attributed to the high wildlife grazing manure deposition, but also the low nitrification potential associated with the mopane woodlands [25]. These soils also showed minute levels of nitrite (NO_2^-). Nitrite, is the intermediate product in the conversion of NH_4^+ to NO_3^- and is toxic to plants; fortunately it is rapidly converted to NO_3^-) or leached in sandy soils. Thus the minute levels of the NO_2^- observed in this study. Nitrate N in these soils were all below the detection level. Although this was partly explained by the low nitrification levels of grassland and mopane soils, it could also be due to the very high sand content of these soils (Table 1). Thus, any NO_3^- that may arise from nitrification is either taken up by plants right away or lost by leaching as sand soils are highly susceptible to leaching [26].

3.3 Soil Microbial Diversity and Biomass

Bacterial populations in the different land uses did not differ significantly except in the grassland which showed lower bacterial counts compared to the other four land uses (Table 3). However, the grassland had significantly higher ($P=.05$) fungal populations than the other land uses. Fungi play a major role in decomposition of organic residues, thus are bound to be higher in grassland than in the woodland surface soils. Mopane woodlands are known to exhibit allelopath behaviour to other plants especially grasses. This is due to its roots' ability to secrete phenolic compounds which inhibit the growth of other plants [6]. Thus, there is usually very low population of grasses or other plants growing under the mopane canopy [26]. In Serongamopane woodlands, very few grass species grow in the tree canopy, as such the low fungal population observed due to lack of the substrates. Actinomycetes populations were highest in the mopane woodland. Mopane leaves are known to contain tannins and other compounds highly resistant to most microbial degradation. Although resistant to other microorganisms, actinomycetes are capable of degrading highly resistant compounds and can carry out the process at high soil temperatures (>40°C). Seronga region has very high temperatures (>40°C) during the dry hot season, as such actinomycetes are most likely the only microorganisms that can survive and degrade resistant compounds at those soil temperatures.

Table 3. Mean microbial populations in the different land uses of Seronga soils

Land use	Bacteria	Fungi	Actinomycetes
		(CFU/g soil)	
Grassland	3.4×10^6	1.6×10^6	6.4×10^5
Woodland Ximenia	1.1×10^7	3.1×10^4	3.3×10^5
Woodland mopane	5.5×10^7	1.4×10^4	9.4×10^5
Cultivated	5.2×10^7	1.2×10^5	1.5×10^4
Fallow	5.3×10^7	1.2×10^5	9.1×10^5

Biophagic protozoa populations in all the land uses were very low. With none observed in all the other land uses except, 70 and 90 MPN/g soil in the cultivated and fallow fields respectively. This is in accordance with the low bacteria populations of these soils. Protozoans have a low grazing population limit of 10^6. Below that they form cysts. Similar studies from mopane soils in the Okavango have also shown insignificant biophagic protozoan populations [25].

Table 4 shows the most probable number (MPN) of nitrifiers in the different land use soils. Most probable number of (NH_4^+) nitrifiers in the different land uses of Seronga showed the lowest levels in the grassland. The paucity of the nitrifiers confirms the low nitrification ability of the grassland. This land use systems also showed significantly higher *(P=.05)* NH_4^+ than the other land use types (Table 2). The NO_2^- nitrifiers on the contrary did not differ with land use except for the woodland Ximenia, a parameter that may be associated with low availability of the substrate in all the land use systems probably due to leaching of nitrites before the bacteria could use them.

Table 4. Most probable number of nitrifiers and microbial biomass C and flush of N in Seronga soil under different land use

Land use	MPN nitrifiers±SD/g soil	
	NH4+	NO2-
Grassland	78±24a	14±5a
Woodland *Ximenia*	170±36b	260±29b
Woodland mopane	330±38c	22±6a
Cultivated	790±54d	22±7a
Fallow	170±29b	28±11a

Fig. 1 also shows the microbial biomass carbon contents of the soils under different land uses. The highest microbial biomass was observed in the grassland (120.76 mgC/kg soil) followed by the cultivated and then soils under fallow. The high microbial biomass observed in the grasslands also corresponds well with the high total carbon observed there (Table 2). This may indicate high substrate for the microorganisms and as such the high microbial biomass observed. On the contrary, mopane soils only had 11.73 mgC/kg soil. The low microbial biomass carbon observed in the mopane soils actually corresponds with the low total carbon (Table 2) and fungal counts (Table 3) observed in those soils. Due to their size, soil fungi contribute more to microbial biomass than bacteria. Mopane soils also have very little other vegetation in terms of grass species growing on them. As such, the A_1 horizon has lower root density compared to the other land uses. Grasslands due to the fibrous root system prevailing in the A_1 horizon habour a lot of microorganisms on their roots thus the higher biomass C observed in grassland and the fallow fields.

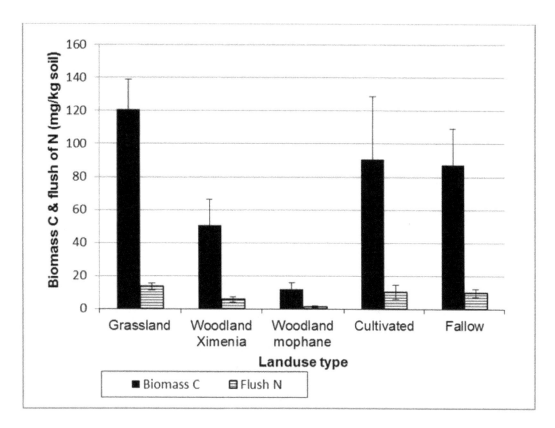

Fig. 1. Influence of land use microbial biomass C and flush of N.
Error bars represent standard deviations.

Crop residues and animal manure amendments in cultivated fields may increase microbial population due to their role in decomposition and mineralisation [26]. The increased microbial population may thus also contribute to increased microbial biomass C and flush of nitrogen observed in cultivated fields as opposed to the two woodland systems.

4. CONCLUSION

Soil microbial population, biomass and soil nitrogen, carbon and sulphur status in the surface soils of the Okavango channel basin of Seronga differed with land use. Overall total N was very low. However, total N and Cwas significantly higher in the grasslands close to the channel and was lowest in the mopane woodlands furthest from the channel. Both cultivated fields and grasslands had significantly higher *(P=.05)* N, C, and microbial biomass carbon than the other land use soils studied. Cultivation of woodlands followed by soil amendments with cattle manure and crop residues seems to increase total N and microbial biomass C. However these decrease when the fields are left fallow. Overall this study indicates that the fallow system practiced lowers the N and C content of the previously cultivated soils, as such is an unjustified practice.

ACKNOWLEDGEMENTS

The authors would like to thank Aone Raleswele and Thabo Hamiwe for assistance in sample analysis and the Okavango Research Institute field staff for assistance in sample collection. The authors also very grateful to Prof. Susan Ringrose for guidance in proposal writing and the German Federal Ministry of Education and Research (*Bundesministerium für Bildung und Forschung)* for funding "The future Okavango, Sp 03" that enabled this work.

COMPETING INTERESTS

Authors have declared that no competing interests exist.

REFERENCES

1. Bonyongo MC, Bredenkamp GJ. Floodplain vegetation in the Nxaraga Lagoon area, Okavango Delta, Botswana. South African Journal of Botany. 2000;66:15-21.
2. Staring GJ. Soils of the Okavango Delta. United Nations Educational, Scientific and Cultural Organisation (UNESCO), Gaborone; 1978.
3. Mengel K, Kirkby EA. Principles of Plant Nutrition. Int. Potash Ins, Bern, Switzerland; 1982.
4. Kraiser T, Gras DE, Gutierrez AG, Gonzalez B, Gutierrez RA. A historical review of nitrogen acquisition of plants. J Exp Botany. 2011;62:1455-1466.
5. Chu H, Fujii, T, Morimoto S, Lin X, Yagi K, Hu J, Zhang J. Community structure of ammonia-oxidizing bacteria under long-term application of mineral fertilizer and organic manure in a sandy loam soil. Applied and Environmental Microbiology. 2007;73(2):485-491.
6. Molotja GM, Ligavha-Mbelengwa MH, Bhat RB. Antifungal activity of root, bark, leaf and soil extracts of *Androstachys johnsonii* P rain. African Journal of Biotechnology. 2011;10(30):5725-5727.
7. Anderson JM, Ingram JSI. Tropical Biology and Fertility: A Handbook of Methods, C.A.B International, UK; 1993.
8. Keeney DR, Nelson DW. Nitrogen-Inorganic forms. In: Page AL, Miller RH, Keeney DR, editors. Methods of soil analysis. Part 2. Chemical and microbiological properties. 2nd ed. Argon Madison, WI; 1982.
9. Williams ST, Wellington EMH. Actinomycetes. In: Page AL, Miller RH, Keeney DR, editors. Methods of soil analysis. Part 2. Chemical and microbiological properties. 2nd ed. Argon Madison, WI; 1982.
10. Gupta SRVV, Germida JJ. Population of predatory protozoa in field soils after 5 years of elemental S fertilizer. Soil Biology and Biochemistry.1988;20:787-791.
11. Schmidt EL, Belser LW. Nitrifying bacteria. In: Page AL, Miller RH, Keeney DR, editors. Methods of soil analysis. Part 2. Chemical and microbiological properties. 2nd ed. Argon Madison, WI; 1982.
12. Cochran WG. Estimates of bacteria densities by means of the "most probable number". Biometrics. 1950;6:105-106.
13. Jenkinson DS, Powlson D. The effects of biocidal treatments on metabolism in the soil. SoilBiology and Biochemistry. 1976;8:167-177
14. Jenkinson DS, Ladd JN. Microbial biomass in soil, measurement and turnover. P415-475. In Paul EA, Ladd JN. Editors. Soil Biochemistry Vol. 5. Marcel Dekker. Inc., New York; 1981.

15. Lindwall CW, Anderson DT Agronomic. Evaluation of minimum tillage systems for summer fallow in southern Alberta. Can. J. Plant Scie. 1981;61:247-253

16. Pettygrove GS, Doane TA, Horwath WE, Wu JJ, Mathews MC, Meyer DM. Mineralization of nitrogen in dairy manure water. Western nutrient management conference. Salt Lake City, UT. 2003;5:36-41.

17. Iwegbue CMA, Emuh FN, Bazunnu AO, Eguavoen O. Mineralization of nitrogen in hydromorphic soils amended with organic wastes. J. Appl. Sci. Environ. Manage. 2011;15(2):257–263.

18. Benka-Coker MO, Ekundayo JA. Effects of an oil spill on soil physico–chemical properties of a spill site in the Niger Delta area of Nigeria. Environmental monitoring assessment. 1995;36:93-104.

19. Cohen E, Okon Y, Kigel J, Nur I, Henis Y. Increase in dry weight and total nitrogen content in zea mays and setariaitalica associated with nitrogen-fixing *Azospirillum* spp. Plant Physiol. 1980;66(4):746-9.

20. Siddiqui AM, Bal AK. Nitrogen fixation in peanut nodules during dark periods and detopped conditions with special reference to lipid bodies. Plant physiology. 1991;95(3):896-899.

21. Omari K, Mubyana T, Matsheka MI, Bonyongo MC, Veenendaal E. Flooding and its influence on diazotroph population and soil nitrogen levels in the Okavango Delta soils. South African Journal of Botany. 2004;70:734-740.

22. Wolkowski RP, Kelling KA, Bundy LG.Nitrogen management on sandy soils, Univ.of Wisconsin-Extension Bull. no. A3634; 1995.

23. Zechmelster-Boltenstern S, Smith KA. Ethylene production and decomposition in soils. Biology Fertility of Soils. 1998;26:354-361.

24. Bonyongo MC, Mubyana T. Soil nutrient status in vegetation communities of the Okavango Delta floodplains. South African Journal of Science.2004;100:337-340

25. Siele MP, Mubyana-John T, Bonyongo MC. The effects of soil cover on soil respiration in the Mopane (*Colophospermum mopane*) woodland of Northwestern Botswana. Dynamic Soil, Dynamic Plant. 2008;2(2):61-68.

26. Britto DT, Kronzucker HJ. NH_4^+ toxicity in higher plants: a critical review. J. Plant Physiology. 2002;159:567-584.

6

Climate Change Diplomacy- Apparatus for Climate Change Mitigation and Adaptation: A Reflection in the Context of Bangladesh

Mohammad Tarikul Islam[1*]

[1]Department of Government and Politics, Jahangirnagar University, Bangladesh.

Author's contribution

The only author performed the whole research work. Author MTI wrote the first draft of the paper. Author MTI read and approved the final manuscript.

ABSTRACT

The objective of this analysis is mainly to examine the current trend of climate change diplomacy carry forward by the Government of Bangladesh. This focus on effectiveness of Climate Change (CC) Diplomacy for the climate change victim nation like Bangladesh. Some interrelated issues addressed here are: (i) essence of climate change diplomacy for the developing nations mostly affected by climate induced disasters; (ii) performance of Bangladesh in the bilateral and multilateral negotiations. The findings of the analysis suggest that, climate change diplomacy of the government of Bangladesh is not formally fashioned yet to influence foreign governments and/or multilateral institutions towards extending highest support in mitigating and adapting the climate change impact. It is apparent that, climate change diplomacy does not seem to have emerged as an integral component of its international relations to complement the national efforts through tie up with international affiliations. The paper is concluded with a roadmap to make the climate change diplomacy more effective so that Bangladesh could avail optimum assistance from the international community/alliance to mitigate and adopt climate change for reduction of vulnerability of the community recurrently affected by climate change induced disasters. With the prominence of climate change diplomacy on the top, Bangladesh should have a strategy of playing a pro-active role at the international level in coming years. Such strategy would help Bangladesh to draw on the global assistance in favor of climate change mitigation and adaptation.

*Corresponding author: Email: tarikul.islam81@gmail.com;

Keywords: Climate Change, Diplomacy, Foreign Relations, Negotiations, Ministry of Foreign Affairs, Bangladesh and United Nations.

DEFINITIONS

1. **Climate change** is any long-term significant change in the "average weather" of a region or of the earth as a whole. Average weather may include average temperature, precipitation and wind patterns. It involves changes in the variability or average state of the atmosphere over durations ranging from decades to millions of years. These changes may be caused by dynamic processes on Earth, external forces including variations in sunlight intensity, and more recently by human activities. Some scientists predict that climate change will cause severe disruptions, such as weather related natural disasters, droughts, and famines, which may lead to enormous loss of life. Global warming between 1.6 and 2.8 degrees Celsius over the next three decades would raise sea levels by half a meter.

2. **Bilateral diplomacy** or Bilateralism consists of the political, economic, or cultural relations between two sovereign states.

3. **Track II diplomacy** is a specific kind of informal diplomacy, in which non-officials (academicscholars, retired civil and military officials, public figures, and social activists) engage in dialogue, with the aim of conflict resolution, or confidence-building. This sort of diplomacy is especially useful after events which can be interpreted in a number of different ways, both parties recognize this fact, and neither side wants to escalate or involve third parties for fear of the situation spiraling out of control.

4. **Multilateral diplomacy** is a term in international relations that refers to multiple countries working in concert/a group/alliance on a given issue.

5. **Cultural diplomacy** is a domain of diplomacy concerned with establishing, developing and sustaining relations with foreign states by way of culture, art and education. It is also a proactive process of external projection in which a nation's institutions, value system and unique cultural personality are promoted at a bilateral and multilateral level.

1. INTRODUCTION

Climate change is one of the greatest challenges of the 21st century as increasing evidence of the impacts of climate change and those human actions are contributing to changes in climate. There is a growing apprehension across the world that attaining the consensus and commitment needed to take action call for positioning climate change in a broader foreign policy context. The perceived goal of traditional foreign policy is to provide stability and security as a foundation for human well-being, global harmony, freedom and prosperity. However, in today's increasingly inter-connected world, the traditional instruments of diplomacy are not always effective in tackling global threats. Established alliances and procedures are hard-pressed to be effective against a threat such as climate change, when the greenhouse gas emission is not the ambition of any one "unreceptive" influence.

In order to address the climate change challenge, it requires new thinking in foreign policy— thinking that considers engagement on climate change not only in the sphere of environment, but also outside the milieu container. Science (climate scientists) and politics (diplomats and Foreign Ministry officials) may not always speak the same language, but climate change diplomacy inevitably brings them together into a "marriage of convenience".

With the intention of addressing the special needs of vulnerable countries like Bangladesh/Maldives, there is consensus between science and politics that the principle of "common but differentiated responsibility" offers the best paradigm and institutional framework to understand and confront the asymmetries in the international system. Although not a significant contributor to climate change, Bangladesh is one of the countries is most at risk from its projected impacts. Climatic events like cyclones, tornadoes and floods have in recent years become less predictable, and more severe and frequent. Bangladesh is one of the most disaster prone countries in the world with great negative consequences being associated with various natural and climate change induced hazards. The geophysical location, land characteristics, multiplicity of rivers, and the monsoon climate render Bangladesh highly vulnerable to natural hazards. The coastal morphology of Bangladesh influences the impact of hazards on the area. Especially, in the south eastern area, natural hazards increase the vulnerability of the coastal dwellers and slow down the process of social and economic development.

Since independence in 1971, the country has endured almost 200 disaster events – cyclones, storm surges, floods, tornadoes, earthquakes, droughts and other calamities – causing more than 500,000 deaths and leaving prolonged damage to quality of life, livelihoods and the economy [1]. Bangladesh is a low-lying deltaic country in South Asia formed by the Ganges, the Bharmaputra and the Meghna rivers. It is a land of about 136.7 million people within its 147,570 sq. km territory. More than 310 rivers and tributaries have made this country a land of rivers. Environmental displacement with the premise of increased frequency of natural disasters and the adverse impacts of climate change, human security seems to be in jeopardy. It is bordered on the west, north and east by India, on the south-east by Myanmar and on the south by the Bay of Bengal.

There is an increasing realization in the international community that achieving the consensus and commitment needed to take action requires positioning climate change in a broader foreign policy context. The ostensible goal of western foreign policy is to provide stability and security as a foundation for human well-being, global freedom and prosperity. However, in today's increasingly inter-connected world, the traditional instruments of diplomacy are not always effective in tackling global threats. Established alliances and procedures are hard-pressed to be effective against a threat such as climate change, when the cause (greenhouse gas emissions) is not the ambition of any one "hostile" power. Addressing the climate change challenge requires new thinking in foreign policy—thinking that considers engagement on climate change not only in the sphere of environment, but also outside the environment box. The country pursues a moderate foreign policy that places heavy reliance on multinational diplomacy, especially at the United Nations. Since independence in 1971, the country has stressed its principle of friendship towards all, malice towards none in dictating its diplomacy.

As a member of the Non-Aligned Movement, Bangladesh has tended to not take sides with major powers. Since the end of the Cold War, the country has pursued better relations with regional neighbors. One of the core principles of the foreign policy of Bangladesh is to uphold the right of every people freely to determine and build up its own social, economic and political system by ways and means of its own free choice. This is a sign of venture of Bangladesh to ensure the right of its climate change affected communities.

The Ministry of Foreign Affairs (MoFA) to the Government of Bangladesh has a major role to play in global Climate Change Diplomacy at the international level in the form of bilateral or multilateral modus. It is already clear that almost every significant bilateral meeting between the foreign minister and her counterparts, such as Hilary Clinton, includes climate change diplomacy as an important topic. The recent ministerial level meeting of the Climate Vulnerable Forum (CVF) hosted by Bangladesh under the auspicious of the Ministry of Environment and Forest (MoEF) and the MoFA speaks about an excellent example of cooperation between these two ministries. Presence of UN Secretary General in the meeting displayed the sincere endeavor of Climate Change Diplomacy by the Government of Bangladesh.

The paper is intended to appraise role of Bangladesh at different steps of negotiation to draw attention of international community to mobilize resources for climate change adaptation in Bangladesh. The study is decorated with content analysis. However, it took place in Dhaka, Bangladesh from July -September 2012 through comprehensive examination of important articles, books and other sources pertaining to the study topic. In doing so, first the study has looked into the role of ministry of foreign affairs of the government of Bangladesh for its climate change negotiations efforts at different level. Simultaneously, an attempt is made to study consequence of climate change in Bangladesh.

2. CLIMATE CHANGE MITIGATION AND ADAPTATION: DEFINITION

Policy responses to climate change are rest with mitigation and adaptation. Mitigation addresses the root causes, by reducing greenhouse gas emissions, while adaptation seeks to lower the risks posed by the consequences of climatic changes. Both approaches will be necessary, because even if emissions are dramatically decreased in the next decade, adaptation will still be needed to deal with the global changes that have already been set in motion.Climate change mitigation is action to decrease the intensity of radioactive forcing in order to reduce the effects of global warming. In contrast, adaptation to global warming involves acting to tolerate the effects of global warming. Most often, climate change mitigation scenarios involve reductions in the concentrations of greenhouse gases, either by reducing their sources or by increasing their sinks.

The UN defines mitigation in the context of climate change, as a human intervention to reduce the sources or enhance the sinks of greenhouse gases. Examples include using fossil fuels more efficiently for industrial processes or electricity generation, switching to renewable energy (solar energy or wind power), improving the insulation of buildings, and expanding forests and other "sinks" to remove greater amounts of carbon dioxide from the atmosphere. Some assert that also non-renewable sources of energy such as nuclear power should be seen as a way of reducing carbon emissions. The International Atomic Energy Agency advocates this approach. However, even while reporting to the UN, the IAEA is independent from it and in no way affiliated with the UNFCCC [2]. Scientific consensus on global warming, together with the precautionary principle and the fear of abrupt climate change is leading to increased effort to develop new technologies and sciences and carefully manage others in an attempt to mitigate global warming. Most means of mitigation appear effective only for preventing further warming, not at reversing existing warming. The Stern Review identifies several ways of mitigating climate change. These include reducing demand for emissions-intensive goods and services, increasing efficiency gains, increasing use and development of low-carbon technologies, and reducing fossil fuel emissions.

3. ESSENCE OF CLIMATE CHANGE DIPLOMACY FOR THE DEVELOPING COUNTRIES: CRITICAL ANALYSIS

Negotiation is a dialogue between two or more people or parties, intended to reach an understanding, resolve point of difference, or gain advantage in outcome of dialogue, to produce an agreement upon courses of action, to bargain for individual or collective advantage, to craft outcomes to satisfy various interests of two people/parties involved in negotiation process. Negotiation is the core part of Climate Change Diplomacy thatensures the involvement of each party for bilateral or multilateral discussion in order to gain an advantage for them by the end of the process. Climate scientists, activists and climate change negotiators often refer to the "overwhelming" scientific evidence of the impact of GHG (Greenhouse Gas) emissions: extreme climatic events (floods, shorter and warmer growing seasons), sea-level rise (causing erosion and salinity in coastal areas), melting of glaciers and many other phenomena [3].

While all of these are shared global problems, their impact will be much more severe in the developing countries exposed to climate change risk. Focusing on the impact of climate change on developing countries, the UN Inter-governmental Panel on Climate Change (IPCC) found in its 2007 Synthesis Report that by mid-century, "climate change is expected to reduce water resources in many small islands, e.g., in the Caribbean and Pacific, to the point where they become insufficient to meet demand during low-rainfall periods". Further, global sea-level rise is "expected to exacerbate inundation, storm surge, erosion and other coastal hazards, thus threatening vital infrastructure, settlements and facilities that support the livelihood of island communities," and "erosion of beaches and coral bleaching is expected to affect local resources" [4]. The report also found that with higher temperatures, "increased invasion of non-native species is expected to occur, particularly on mid- and high-latitude islands". Corroborating most of these findings, the UNFCC Secretariat, in the 2008 Climate Change: Impacts, Vulnerabilities and Adaptation in Developing Countries, identified that "all Caribbean, Indian Ocean and North and South Pacific small island states will experience warming". In an attempt to address the fundamental concerns, negotiation across level has emerged as an impending and worthwhile force. Climate change diplomacy results in promotion of mitigation and adaptation of climate change impact. For instance, at the Durban negotiations, countries focused on three steps to ensure the developed world can meet its agreed responsibilities: establish funding sources based on international trade; define annual targets for the scale-up; and adopt a transparent, centralized accounting system.

In order to be sustainable, climate change diplomacy must address the technological, financial and policy needs of SIDS in pragmatic ways. This is not a matter of "charity" or "aid". It is an obligation owed to them by the international community as a whole. While global resources are never in short supply to achieve this, only a fair, equitable and distributive multilateral governance facility stands to protect and promote their needs and meet their expectations as vulnerable societies in a global village characterized by asymmetries and socio-economic inequalities between nation-States. The gravity of climate change diplomacy is rest with negotiations with different actors assertively to get the best out of the international provision responding to the mitigation and adaptation for the climate change victim developing nation. Climate Change Diplomacy envisages confidence building amongst different parties across level so as to ensure that, climate change victim is duly supported in all aspects for mitigation and adaptation.

It is now apparent that negotiators are rethinking to take "precautionary" and "adequate" measures to anticipate climate change, prevent or minimize its causes and reduce its adverse effects. This means taking action even amid scientific uncertainty. The types and sources of funds are also an issue. It is not clear what proportion of adaptation funding will be pure grants, loans with concessionary terms, or purely market-rate loans. Vulnerable countries are not able to repay loans for adaptation, nor should they have to. In addition, the Cancun texts promise 'predictable' funds, which is essential for developing countries to budget and plan for adaptation responsibly. But predictability has not increased since the 2009 meeting in Copenhagen, as wealthy governments have not mustered the will, political support or taxes to raise climate finance. "Scaled up" is another phrase that has not been adequately addressed [5]. After years of watching wealthy nations put token voluntary contributions into UN climate funds, developing nations pushed for meaningful, scaled-up funding at Copenhagen. Copenhagen and Cancun also promise "new and additional" funding. These much-debated words suggest climate finance will be over and above conventional development aid known as Official Development Assistance (ODA) - but their meaning in practice has been ambiguous.

Most donor countries have similarly failed to justify the way they divide the burden of confronting climate change. A recent study found that only two of 10 contributors that reported their fast-start finance activities to the UNFCCC indicated how they calculated their fair share of funding. The Copenhagen Accord and Cancun Agreements promised balanced allocation between adaptation and mitigation. Fast-start donors have pledged between $4.8 billion (3.59 billion euros) and $6.3 billion to adaptation, 19-25 percent of total climate finance - only a small rise on the 11-15 percent pledged a year ago [6]. Adaptation funds should go first to those most vulnerable to climate impact, as promised in several recent agreements, including Cancun. Vulnerable groups are not only geographically exposed to physical threats such as sea level rise, drought or disease, but are especially susceptible to harm because of poverty and powerlessness.

Despite pledges of transparency in Bali, climate finance has been poorly reported and impossible to track and verify. Climate finance is highly fragmented, with dozens of donors, including governments, multilateral agencies, private foundations and civil society organizations. With so many funding channels and very little information, it is difficult for both donors and recipients to assess where money is going. Developing countries are left not knowing how much support to expect, when and for what. There are three essential steps Durban delegates have directed efforts together to take toward robust, effective adaptation funding that fulfills past promises. Climate Change Diplomacy loose ends as an alternative and viable force in order to tackle climate change impact by achieving the following goals:

To fund the scale-up period and beyond, negotiators from climate victim nations should work out a series of financing mechanisms that are international, constant and substantial in size. Climate finance negotiations have a blind spot: the scale-up period from 2013 to 2019 [7]. In this period - after the fast-start years but before the $100 billion-per-year pledge for 2020 - developed nations need defined targets for each year and mechanisms to keep the expansion of funding on track. Only then will they develop systems capable of generating the amounts committed by 2020. Notwithstanding the creation of United Nations Framework Convention on Climate Change (UNFCCC)-led funds, most money in the next few years will likely flow bilaterally or through multilateral channels not governed by the convention. This makes transparency and central accounting even more crucial. Durban negotiators is one of the evidences to press the need for creating a central accounting framework and registry, perhaps under the UNFCCC's Standing Committee; provide a global definition of "new and additional" adaptation finance; and agree to standardize a format for more precise project-level reporting of financial flows.

To envisage the future climate change diplomacy, it is pivotal that any agreement going forward specify and deliver on fair and effective funding, as was promised in Copenhagen and Cancun. The funding must be adequate and predictable, and be delivered justly and transparently. Poor and vulnerable nations should be the first to receive funds, and should have a say in fund governance.

4. WORLD CLIMATE CONFERENCES: COMPLIMENTARY TO CLIMATE CHANGE DIPLOMACY

The principle of "common but differentiated responsibilities" recognizes the asymmetries of the international system, especially the differential levels of technological, financial, economic and human capacities between industrialized/developed and developing countries in international environmental negotiations. Despite these asymmetries, every nation has an obligation to participate in joint efforts to tackle shared global environmental problems according to each nation's capacity and level of development. However, industrialized countries have an obligation to bear a greater burden of these shared problems. The First World Climate Conference was held on 12-23 February 1979 in Geneva and sponsored by the World Meteorological Organization (WMO). It was one of the first major international meetings on climate change. Essentially a scientific conference, it was attended by scientists from a wide range of disciplines. The Conference led to the establishment of the World Climate Programme and the World Climate Research Programme. It also led to the creation of the Intergovernmental Panel on Climate Change (IPCC) by WMO and UNEP in 1988 [8].

The Second Climate Conference was held on 29 October to 7 November 1990, again in Geneva. It was an important step towards a global climate treaty and somewhat more political than the first conference. The main task of the conference was to review the WCP set up by the first conference. The IPCC first assessment report had been completed in time for this conference. The scientists and technology experts at the conference issued a strong statement highlighting the risk of climate change. The conference issued a Ministerial Declaration only after hard bargaining over a number of difficult issues; the declaration disappointed many of the participating scientists as well as some observers because it did not offer a high level of commitment. Eventually, however, developments at the conference led to the establishment of the United Nations Framework Convention on Climate Change (UNFCC), of which the Kyoto Protocol is a part, and to the establishment of the Global Climate Observing System (GCOS), a global observing system of systems for climate and climate-related observations. World Climate Conference-3 (WCC-3) was held in Geneva, Switzerland, 31 August - 4 September 2009 [9]. Its focus was on climate predictions and information for decision-making at the seasonal to multi-decadal timescales. The goal was to create a global framework that will link scientific advances in these climate predictions and the needs of their users for decision-making to better cope with changing conditions.

The Conference also aimed to increase commitment to, and advancements in, climate observations and monitoring to better provide climate information and services worldwide that will improve public safety and well-being. WCC-3 outcomes also intended to contribute to the achievement of the United Nations Millennium Development Goals and broader UN climate goals, including the Hyogo Framework for Action on Disaster Risk Reduction. The

Conference theme complemented global work under way to help societies adapt to climate change in line with Bali Action Plan, especially the Nairobi Work Programme.

4.1 Role of United Nations in Climate Change Diplomacy

At the core of international efforts to address climate change are the United Nations Framework Convention on Climate Change and its Kyoto Protocol. These two treaties represent the international response so far to the compelling evidence, compiled and repeatedly confirmed by the Intergovernmental Panel on Climate Change, that climate change is occurring, and that it is largely due to human activities. Countries agreed on the Convention on 9 May 1992, and it entered into force on 21 March, 1994. But even as they adopted the Convention, however, governments were aware that its provisions would not be sufficient to adequately address climate change. At the first Conference of the Parties, held in Berlin, Germany in early 1995, a new round of talks was launched to discuss firmer, more detailed commitments. After two and a half years of intensive negotiations, a substantial extension to the Convention was adopted in Kyoto, Japan in December 1997 [10]. This Kyoto Protocol established legally binding emissions targets for industrialized countries, and created innovative mechanisms to assist these countries in meeting these targets.

The Kyoto Protocol entered into force on 18 November 2004, after 55 Parties to the Convention had ratified it, including enough industrialized countries - who have specific targets - to encompass 55 per cent of that group's carbon dioxide emissions in 1990. The debate surrounding climate change on future severity, how much is man caused, and what the solutions might be, has been becoming increasingly vigorous with data and reports providing concrete evidence for global warming. The Intergovernmental Panel on Climate Change (IPCC), was established by the World Meteorological Organization (WMO) and the United Nations Environment Programme (UNEP) to assess scientific, technical, and socio-economic information relevant for the understanding of climate change, its potential impacts and options for adaptation and mitigation. It recently finalized its Fourth Assessment Report: Climate Change 2007. In October 2007 the IPCC and Albert Arnold (Al) Gore shared the Nobel Peace Prize. The UN Framework Convention on Climate Change (UNFCCC) presents the appropriate forum to do this. It has been expanded by the Kyoto Protocol which includes emission reduction commitments for developed countries over the period 2008-2012 [11]. A new international climate change deal must be put in place in time to ensure that necessary action is undertaken immediately after 2012 when the current phase of the Kyoto Protocol ends. Therefore, comprehensive negotiations on a new climate deal need to begin without further delay.

The persistent United Nations Climate Change Conference is the biggest global forum of its member states, NGOs, CSOs and individuals to press the need for mitigation and adaptation of climate change adopting new policies/ road map unanimously. The Intergovernmental Panel on Climate Change (IPCC) warned in November 2011 that extreme weather will strike as climate change takes hold. Heavier rainfall, storms and droughts can cost billions and destroy lives. Estimates suggest that every dollar invested in adaptation to climate change could save $60 in damages.NGOs and civil society have long provided their expertise and advocated for reform with regards to issues such as climate change, and the environment, to multilateral institutions and national governments. These efforts take a variety of forms, from advocacy to grassroots level education and action, to sharing experience and knowledge with policymakers, to publishing position papers, and to taking part in conferences surrounding these issues.

Since the 1972 United Nations Conference on the Human Environment, an increasing number of NGOs and other members of civil society from all around the globe have participated in international conferences on the environment. Civil Society representatives have brought invaluable expertise and intervention strategies to international meetings. The 18th Conference of the Parties to the United Nations Framework Convention on Climate Change in Doha, Qatar ended with lot of promises to address climate change [12]. This is the first time that the UN climate change conference has been held in a Gulf country. Although developing countries have warned those refusing to join the second commitment period of the Kyoto Protocol, they cannot enjoy the benefits of market mechanisms in the Kyoto Protocol, such as the clean development mechanism. Those countries not agreeing to the second commitment period of the Kyoto Protocol are happy not to make commitments and set emissions limits.

5. PROSPECTS OF CLIMATE CHANGE DIPLOMACY OF BANGLADESH: AN OVERVIEW

Bangladesh is one of the most vulnerable countries to climate change due to global warming. According to current scientific understanding, the state of well-being and survival of the people in Bangladesh will be under serious threat from climate change over the coming decades. Being associated with various natural and climate change induced hazards, Bangladesh has been the worst victim in the world as would be evident from Fig. 1 as well as Fig. 2.

- One third of population below the poverty line and 17% or some 27 million people still live in extreme poverty
- Sea level rise has the potentials to displace nearly 30 million people living in the coast
- in terms of people exposed to Bangladesh is ranked globally:
  - 1st for floods, 3rd for tsunamis and 6th for cyclones
    - 14% GDP exposed to disasters per year – the highest ranking in the world
    - Between 1980-2008:
  - 219 natural disasters
  - more than seven disasters per year
  - causing over USD 16 billion in damage
    - 93% river flows coming across border

Fig. 1. Disaster and climate profile of Bangladesh
Source: United Nations Development Programme, Climate Change, Environment and Disaster profile of Bangladesh, 2012

As a country most vulnerable to climate change, Bangladesh has been experiencing sufferings caused by climate change. the National Adaptation Programme for Action (NAPA) highlighted prediction on changing pattern of temperature, rainfall and sea level rise in Bangladesh due to climate change impact (Fig. 3).

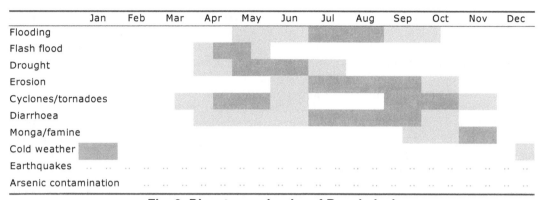

Fig. 2. Disasters calendar of Bangladesh
Source: World Food Programme Bangladesh, 2011

Year	Temperature change (°C) mean	Rainfall change (%) mean	Sea level rise (cm)
2030	1.0	5	14
2050	1.4	6	32
2100	2.4	10	88

Fig. 3. Statistical data of predication for Bangladesh due to climate change impact
Source: National Adaptation Programme for Action, Department of Environment, Government of Bangladesh, 2005

The above projections of climate parameters may be concluded that the country will be highly susceptible to: (a) increased flooding, both in terms of extent and frequency; (b) increased moisture stress during dry periods leading to increased drought susceptibility in terms of both intensity and frequency; and (c) increased salinity intrusion during the low flow conditions. These changes in the physical system of the country will directly affect a number of major productive systems that include (a) crop agriculture, (b) livestock production, (c) aquaculture and fish production, (d) coastal shrimp production, and (e) forest and vegetation and (f) livelihoods of poor and marginal households. Due to changes in temperature and humidity, human health will also be affected. The high susceptibility to water-based natural hazards will affect settlement of the population and also physical immobile infrastructure. Based on secondary sources, the following sub-sections provide brief understanding on anticipated impacts of climate change on bio-physical aspects of the country.

As a country most vulnerable to climate change, Bangladesh has been experiencing sufferings caused by climate change. Bangladesh's endeavor towards making best use of diplomatic affiliation in an attempt to address the issue of climate change is noticeable [13]. For example, China fully understands and respects the concerns of Bangladesh over it. During international negotiations on climate change, China has always been supporting the legitimate and reasonable requests of Bangladesh and the Least Developed Countries as a whole. As developing countries, China, India, Malaysia and Bangladesh should stick to the principle of "common but differentiated responsibilities" and work together to safeguard the common interests of developing countries. China and Bangladesh are both victims of climate change. On this issue, the two countries face the same challenges and our basic interests are the same.

Over the years, China has already carried out cooperation with Bangladesh in the area of adaptation. For example, China helped Bangladesh in projects of river dredging. The Chinese Government provided relevant training for Bangladeshi officials and technicians. Bangladesh should attempt to strengthen the cooperation with Bangladesh in this regard on the basis of "equal consultation, mutual benefit and common development. It is apparent that, Bangladesh government deliberately presents itself "as a worst victim, peace loving, and responsible actor" and "as a poor developing country." Both discourses are designed to accomplish the broader diplomatic agendas. The rhetoric of being "a responsible actor" enables Bangladesh to raise its international profile and pursue its interests more easily; and that of being "a poor developing country" underlines its need for foreign assistance to tackle the adverse impact of climate change.

In fact, the negotiations are for solving a global problem. Therefore, the world needs to work collectively. Another thing is the real negotiations happen behind closed doors, in corridors, near the swimming pools and places like that. However, Bangladesh, over time, has learned a lot. It has played a proactive role in several global negotiations. Negotiation of Bangladesh Delegates in the conference of parties (COP) of the United Nations Framework Convention on Climate Change (UNFCCC) was effective as they strongly presented the real case as the worst climate change victim nation. Bangladesh found to be in positive effort to play a leading role in the last COP. However, first it is useful to assess, dispassionately, the role that Bangladesh has been able to play so far, in order to see how this can be built on and improved going forward.

Role of Bangladesh in the recent Climate Change Conference held in Doha (November 2012) is considerable. Bangladesh had raised strong demands for paying compensation to the climate vulnerable countries as they are not responsible to the global warming [14]. As CVF chairperson, Bangladesh played vital role in negotiation with developed countries as to reduce the carbon emission in order to keep their pledges which speaks of Bangladesh's prospective attachment in the climate negotiations. At the conference, developing countries have asked developed countries to come out with a roadmap to show how the Green Climate Fund will be distributed between 2013 and 2020, and the fund should reach $100 billion by 2020.

The Long-Term Cooperative Action deals with long-term issues such as the pledges of emissions reductions various countries have made and implementation of the goals that should be included in new climate change treaties after 2020. Long-term emissions reduction goals and negotiation results based on the Action should be further implemented in the Durban Platform for Enhanced Action and in future treaties. Though all the participants showed a smile of understanding, they are fully aware that it will not be an easy task to achieve the purpose as originally envisaged. To this end, all contracting parties should engage in inclusive consultations and negotiations. Assessment of Bangladesh's role so far would be to say that Bangladesh certainly played a "prominent "role but far from a "leading" role. They are very active in highlighting Bangladesh's position as a vulnerable country as well as the action that the country is undertaking on tackling climate change. However, banging one's own barrel (however, effectively) does not commend a country to its allies within the LDC Group or other vulnerable countries and it is felt that Bangladesh is only promoting itself. In order to be a leader, a country Bangladesh has to earn the trust of others, who will acknowledge them as leaders. Unfortunately, by promoting itself the opportunity to gain the trust of others was challenged. Hence, for playing a prominent role, the country should able to play a truly leadership role in global forum like COP.

For discharging role of negotiating on country's behalf on climate change, Bangladesh should think about appointment of the high level Climate Change Envoys. Climate change will remain central for Bangladesh for many years) to come and investing in such a High Level Special Envoy who has the trust of the government and the requisite diplomatic (rather than technical) background and skills, along with a team from the relevant ministries, is essential even now. It will also have to take a more nuanced and leading role on behalf of the vulnerable countries. An important element of the strategy going forward for these vulnerable countries is to no longer highlight their vulnerability alone but rather trumpet their actions (both on adaptation as well as on mitigation or low Carbon growth).

This is where they can gain the moral high ground and shame other countries (both the developed countries as well as the large developing countries who have become major polluters now) into action. With regard to gaining financial support from any new international adaptation funds, Bangladesh needs to make the case for getting its due share based on its performance and ability to do the right things with transparency and good governance, rather than staking an a priori claim for a certain share. By proving that Bangladesh can use money well, it will get more than its fair share without having to stake a claim to it in competition with others. Since climate change is an important issue for Bangladesh, it should have a permanent climate change negotiator, who should be a senior or retired ambassador having required skill and knowledge of diplomacy.

Bangladesh is known as one of the most vulnerable countries. We need to make a positive impression across the world. Bangladesh is, however, one of the first countries that prepared a climate strategy and action plan and formed a climate trust fund. Bangladesh has not waited for other countries' money for the fund. Now the challenge is using the fund as everybody is looking at Bangladesh. If Bangladesh does not spend the money well, it will spoil its reputation. It is a problem that the vulnerable countries are asking for funds. Yes, they need more money. But merely asking for money is not enough. The private sector has a tremendous role to play in reducing people's suffering at local, national and global levels. At the global level, it is said that the biggest polluters are the energy, transport and industrial and other companies, not the states. The private sector needs to change the way they run the businesses. They need to go for clean energy.

Bangladesh has good laws for environmental protections, yet it does not have a good track record on enforcing these laws. Sometime the laws are broken with the help of the government people. I have read an article that a big building is constructed on a wetland at the heart of Dhaka. These things are unfortunate. We have to have better enforcement of the laws. We also need to develop a productive relationship between the industries and the citizens and the government, who are the right people to say what has gone wrong. Doing whatever you wish, because you are powerful, is just undermining your own country and your children. Bangladesh's position in the global talks is very close to the least developed countries.

We have a strong bond because we produce the least emission of greenhouse gases yet we are the most vulnerable to climate change. Significant amount of emissions are released by the developed and the fast growing developing countries Brazil, South Africa, India and China. We need to persuade the rich and the fast growing countries realize that they have the right to develop, but they do not have any right to pollute. We must persuade them that there is technology for development keeping the level of emission low. Yes, these technologies are expensive. But they can afford it. Bangladesh foreign policy stands primarily on two pillars: security and development. Foreign policy covers the entire gamut of foreign relations in such areas as, security, trade, manpower export, foreign direct investment, foreign aid, cultural matters, curbing terrorism, humanitarian, and environmental issues.

Foreign policy is no more confined to traditional diplomacy. Foreign policy includes economic and environmental diplomacy. The Research and Evaluation Division will conduct an in-depth research of regional and global events and its anticipated impact on Bangladesh. It will provide the government short-term and long-term policy options within which Bangladesh may likely to operate in 5 or 10 or 15 years [15]. Climate Change Diplomacy is also characterized by complex linkages between foreign and domestic policy and politics the connection between foreign and domestic policy inherent in environmental diplomacy bring new actors to the fore successful environmental diplomacy requires a cooperative, multilateral approach healthy competition for the mantle of international environmental leadership is needed to reinvigorate global environmental diplomacy. Environmental issues are increasingly intertwined with other more traditional areas of foreign relations, including trade and investment, development and human rights and even military security.

Climate Change Diplomacy tries to influence foreign governments and/or multilateral institutions towards certain policies and Climate Change Diplomacy seems clearly to have emerged as an integral component of international relations to complement the national efforts through tie up with international affiliations in mitigating and adapting the climate change. Climate change is growing in importance as a significant new arena of global diplomacy at the very highest levels. As a developing country that is particularly vulnerable to the adverse impacts of climate change this presents a challenge for Bangladesh. At the same time, as the country gains in knowledge about the issue and starts to tackle it in earnest, it also represents an opportunity for it to play a leading role in the international diplomatic arena as well. In order to make the most of such opportunities different ministries of the government will need to enhance their capacities on the issue of climate change diplomacy. Some suggestions for action are described below.

As the lead ministry dealing with climate change the Ministry of Environment and Forests (MoEF) has been playing a leading role on behalf of the country at the meetings of the United Nations Framework Convention on Climate Change (UNFCCC) and the annual Conference of Parties (COP), including the recent COP17 held in Durban, South Africa. Over the years the minister and officials of the ministry as well as expert advisers have gained considerable knowledge of and expertise in the different negotiating tacks and Bangladesh has been playing a leading role at the COPs within the Least Developed Countries (LDC) group to which it belongs. Bangladesh has an opportunity to take over as chair of the LDC group from next year as the chairmanship will move from Africa to Asia at COP18 held in Doha, Qatar in December 2012.This gives Bangladesh a year to lobby amongst the Asian LDCs to gain the chairmanship of the LDC Group.

In fact, negotiations trigger solving a global problem. Therefore, the world needs to work collectively. Another thing is the real negotiations happen behind closed doors, in corridors, near the swimming pools and places like that. However, Bangladesh, over time, has learned a lot in the area of negotiations to address the issue of climate change and therefore, it has been playing a proactive role at several global negotiations since the beginning of 21st century. Negotiation of Bangladesh delegates in the conference of parties (COP) of the United Nations framework convention on climate change was effective as they strongly presented the real case as the worst climate change victim nation. Bangladesh found to be in positive effort to play a leading role in the last COP. However, first it is useful to assess, dispassionately, the role that Bangladesh has been able to play so far, in order to see how this can be built on and improved going forward. Case Study-1 below shows signs of negotiations skill of Bangladesh demonstrated for climate change adaptation in international arena.

Case Study 1. Bangladesh chair Climate Vulnerable Forum

Bangladesh has been accredited for demonstrating leadership to bring all climate victim nations under a single platform. As part of such management, a two-day international conference of the Climate Vulnerable Forum (CVF) held on 13th November 2011, aiming to reach a consensus on various climate issues to work together at the upcoming COP-17 conference in Durban. Ministers and representatives of 33 member countries and observers from 26 countries took part in a discussion. The Dhaka meeting of the Climate Vulnerable Forum (CVF) was considered significant ahead of the COP-17 conference held in Durban from November 28 to December 9, 2011.Prime Minister of Bangladesh and UN Secretary General the inaugural session. The participating countries were Antigua and Barbuda, Bangladesh (Incoming Chair), Barbados, Bhutan, Costa Rica, Ethiopia, Ghana, Grenada, Guyana, Kenya, Kiribati (Present Chair), Liberia, Maldives (First Chair), Marshall Islands, Micronesia, Nepal, the Philippines, Rwanda, Saint Lucia, Samoa, Solomon Islands, Tanzania, Timor-Leste, Tuvalu, Vanuatu and Vietnam. The CVF was formed in 2009 at the initiative of the Maldives to make a broad-based platform of the most vulnerable countries to realize their common goals. A Dhaka Declaration and a Roadmap of Activities was announced and worked as means of negotiations to the Durban COP Conference.

As international finance for climate change from global to national level begins to flow in earnest, the Ministry of Finance, and particularly the Economic Relations Division (ERD) will need to enhance its knowledge of climate change finance, which is different from Official Development Assistance (ODA) with which they have traditionally been familiar. One significant difference between ODA and climate change finance is that ODA is given by developed countries to developing countries under a paradigm of "charity" (or "solidarity") while climate finance is under a treaty obligation under the paradigm of "polluter pays." Thus, the relationship of Bangladesh's officials when dealing with their counterparts from the same developed countries needs to be very different when discussing ODA (where Bangladesh has to accept what is offered on the terms on which it is offered) from discussing climate finance (where Bangladesh can dictate some of the terms). A good example is the position of the LDCs that only grants are acceptable and not loans for climate finance.

The foreign ministry needs to provide regular briefings on climate change diplomacy to its missions abroad and in the longer term should send some junior officers for higher studies in climate change diplomacy. The prime minister is already finding that climate change is a regular topic on the agenda of her meetings with other heads of state and she is quite knowledgeable on the topic. However, given the importance of this topic for Bangladesh in future, she should consider appointing a personal "Special Climate Change Envoy/Adviser" to represent her at important high level international meetings.

A number of both developed as well as developing countries have appointed such "special climate change envoys/advisers." The skills needed for such a position are those of diplomacy, especially within the UN system, rather than scientific expertise. So a current, or former, senior diplomat who has been posted to either New York or Geneva would be a suitable candidate for such a position. The government of Bangladesh should equally reiterate the lexicon of harmony in the international climate policy negotiations. By using metaphors with positive connotations, it sets out to consolidate solidarity and togetherness with the developing world. "Our homeland" and "Mother Earth" in particular can be interpreted by the global community as universal values that "play an important role in argumentation because they allow us to present specific values, those upon which specific groups reach agreement, as more determined aspects of these universal values".

The international community must act in accordance with the principles and provisions of the UNFCCC and its Kyoto Protocol. It must be aware of that helping others is helping oneself and harming others is harming oneself, actively tackle climate change, and work together to make our homeland a better place". Bangladesh's participation in the multilateral dialogues has been shown to be an indicative of its integration into the global society. Bangladesh badly needs the financial and technological support of the West to adopt with the change posed due to climate change. Rhetorically, Bangladesh finds it more feasible to speak of the needs of all developing nations as a group rather than its own interests.

Bangladesh should enhance bilateral relations with neighboring countries not want to be left alone in the negotiations and therefore, should tie up with regional alliance to raise strong voice to draw the attention for resource mobilization for CC adaptation. Nevertheless, Bangladesh should maintain its favorable image as a cooperative partner of the international community. The fear of losing face may constrain its behaviors. A policy cell could be developed by MoFA who will be responsible for Climate Change Diplomacy. MoFA can train diplomats in its Missions abroad who can play a vital role in raising international awareness to the environmental problems of Bangladesh. It should be emphasized that by focusing on environmental issues we are not asking for more aid or charity but rather for international recognition of the environmental importance of Bangladesh. It is a technical and knowledge-intensive affair and all delegations to international negotiations/conferences should be formed accordingly. Delegation members should be selected with appropriate background, continuity in participation and institutional expertise/memory developed over time.

Multilateral diplomacy is not just important but engrained in the conduct of Bangladesh's foreign policy since the country gained its independence. Bangladesh has been working actively towards a global political and socio-economic stability and security within the multilateral system. Bangladesh will therefore promote security, international law as well as development through its active participation in the international fora, especially the United Nations system and its specialized agencies. Climate change diplomacy plays a tremendous role in building consensus, resource mobilization for climate change adaptation and mitigation. The following liner flow chart (Figure 4, developed by author himself) speaks about end result (climate change adaptation and mitigation) of climate change diplomacy.

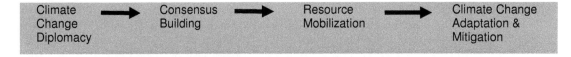

Fig. 4. Liner flow chart shows consequences of Climate change diplomacy

The United Nations through its universal membership and its Charter occupies the central and indispensable role within a multilateral system of governance. Bangladesh recognizes the need to address the pressing social and economic needs of the developing countries and as such looks to the United Nations to advance the development agenda related to underdevelopment and the eradication of poverty. Through Bangladesh's participation in the multilateral fora, Bangladesh has to be vocal with the belief that resolution of international conflict should be peaceful and in accordance with the United Nations and international law. Bangladesh is supporting all initiatives aimed at strengthening the UN and multilateralism. Bangladesh has to play a prominent role in advancing the development agenda of the South through its leadership roles in the NAM and OIC. Diagram in Figure 5 (developed by the author himself) could be considered as an option to promote Climate change diplomacy as the new dimension of diplomacy could be flourished in the form of bilateral diplomacy, track-II diplomacy, multilateral diplomacy as well as cultural diplomacy.

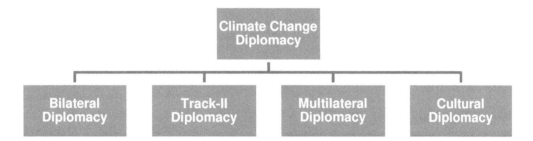

Fig. 5. Different means of Climate change diplomacy

On the top, a road map is proposed for Bangladesh to applause diplomatic negotiations in a win-win agreement. In order to show up the road map, a set of strategies and tactics that are designed to get the country like Bangladesh to the best possible outcome following four steps i.e. defining goals and objectives; promoting research and educate the officials of MoFA about the prevailing global focus, the competition and the other party; defining strategies; as well as defining tactics. In addition to these, MoFA should focus on the following way forwards for outcome-oriented Climate Change Diplomacy:

1. Focus on the problem at hand rather than the people. The goal here is to preserve relationships with the other party (bilateral discussion with other government or between government and international organizations/alliance).
2. Seek to understand the reasons behind any bargaining position (also, look for hidden agenda and personal motivators).
3. Brainstorm alternative agreements and options (track-II diplomacy or cultural diplomacy for unfolding the common interest of two nations)
4. Use objective criteria (pro-active and strong vocal in the international negotiation platforms gaining confidence and support of other climate change victim nations)
5. Know what your best alternatives to the initial negotiation objectives are. (showcase the worst impact of climate change in the international community along with adaptation capacity of the government and the people of Bangladesh)

6. CONCLUSION

It is apparent that, the countries that are most threatened by climate change, like the Bangladesh, must do all they can to break the deadlock. To stress the urgency of arriving at a serious global agreement soon, the governments of Bangladesh can engage in what might be called "anti-diplomacy." given the fact that climate change has become a national security issue, our governments must act in the same manner that they would respond to a national security threat. it is often forgotten how critical a role a single individual, who is both knowledgeable and respected by others, can play at the international level, even if the country he or she comes from is not the biggest or richest. If the Prime Minister and the government are willing to be proactive on climate change diplomacy issue it is quite possible that in a few years' time Bangladesh may also produce such a universally respected individual in climate change diplomacy at the global level. Bangladesh certainly has the potential capability of playing a leading role amongst the fellow-vulnerable countries on the issue of climate change if it plays to its strengths which include capable officials, ministers and a committed prime minister as well as many experts and NGOs who are involved both at home as well as well-networked internationally. With the prominence of Climate change diplomacy on the top of mind, Bangladesh should have a strategy of playing a pro-active role at the international level in coming years to reduce the sufferings of climate change induced disaster victims drawing the global assistance in the area of mitigation and adaptation.

ACKNOWLEDGEMENTS

I sincerely acknowledge the contribution of Dr. Ali Riaz, Distinguished Professor and the Chair, Department of Politics and Government, Illinois State University, USA who guided me in developing this article. I also take the opportunity to express appreciation to my teacher Dr. Bashir Ahmed, Associate Professor, Department of Government and Politics, University of Jahangirnagar, Bangladesh as well as My Wife Ms.Tanzila Mustary, Masters in Political Science, University of Dhaka, Bangladesh as both of them inspired and supported me for presenting the essence of climate change diplomacy for the worst climate change victim nation Bangladesh.

COMPETING INTERESTS

Author has declared that no competing interests exist.

REFERENCES

1.   Asian Development Bank. Climate Change in Asia. Manila. 1994;6-8.
2.   Department of Environment. State of Climate Change in Bangladesh. Dhaka. 2012:29-30.
3.   Department of Disaster Management. Cyclone Emergency Preparedness Plan, GoB. 2012:21-22.
4.   Aginam, Obijiofor.Climate Change Diplomacy and Small Island developing states. UNU. 2012;4:8-13.
5.   Government of Bangladesh. Climate Change Adaptation in Bangladesh, Department of Environment. GoB. 2013;3:7-9.
6.   Asiatic Society. Asia's next challenge: securing the region's water future, Dhaka. 209:56-58.
7.   GED. Poverty, Environment and Climate Mainstreaming. Government of Bangladesh. 2009;24:39-40.
8.   Department of Disaster Management. Annual Report on Disaster Response and Recovery. GoB. 2012;2:36-37.

9. GEF Evaluation Office and Ministry of Foreign Affairs. Evaluation of the operation of the Least Developed Countries Fund for adaptation to climate change. Evaluation Department of the Government of Denmark. 2009;12:33-35.

10. Hedger, M. Climate Finance in Bangladesh: Lessons for Development Cooperation and Climate Finance at National Level. Institute for Development Studies. UK. 2011;3:6-7.

11. Huq, Saleemul. Climate change policy negotiations: can Bangladesh play a leading role. The Daily Star. 2011:8-11.

12. IUCN. Livelihoods and Climate Change in Bangladesh. Dhaka. 2003;7:13-14.

13. UNDP. Climate Change and Development in Bangladesh. Dhaka. 2012;17:24-26.

14. Ministry of Environment and Forests. Bangladesh Climate Change Strategy and Action Plan. GoB. 2013;25:36-38.

15. UNDP. Human Development Report on fighting Climate Change: human solidarity in a Divided World. New York. 2008;11:25-27.

7

Greenhouse Gas Emission Determinants in Nigeria: Implications for Trade, Climate Change Mitigation and Adaptation Policies

A. I. Achike[1] and Anthony O. Onoja[2*]

[1]Department of Agricultural Economics, University of Nigeria, Nsukka, Nigeria.
[2]Department of Agricultural Economics and Extension, University of Port Harcourt, Nigeria.

Authors' contributions

This work was carried out in collaboration between both authors. Author AOO designed the study, performed the statistical analysis, wrote the protocol and wrote the first draft of the manuscript. Both authors read and approved the final manuscript.

ABSTRACT

This study investigated and analyzed the determinants of Carbon Dioxide (CO_2) emission in Nigeria. The study relied on secondary data from World Bank and Central Bank of Nigeria covering 40 years (1970-2009). The data were analyzed using Zellner's Seemingly Unrelated Regression (SURE) model. The results of the analysis show that fossil energy demand or consumption, rents from forestry trade, agricultural land area expansion and farm technology were significant determinants of greenhouse gas (GHG) emission in the study area. On the other hand, the second equation indicated that fossil fuel energy demand was exogenously determined by economic growth rate (proxied by GDP growth rate) and farm technology applied in the country. It was recommended that Nigeria should put in place policies that will tax companies or firms emitting GHGs and utilize such tax proceeds for research and building the capacities of farmers to adapt to deleterious effect of climate change in the country and continent. The development of existing and new technologies for adapting to climate change and variability, building of environmental consciousness of Nigerians through curriculum restructuring and provision of weather information services by the Nigerian governments and their agencies to enable farmers plan against weather uncertainty and risks were also recommended.

*Corresponding author: Email: tonyojonimi@gmail.com;

Keywords: Greenhouse gas emission; fossil energy; economic growth; climate change; Zellner's Seemingly Unrelated Regression (SURE).

1. INTRODUCTION

The world's greatest environmental challenges are climate change and resource depletion [1]. The effects of these challenges and vulnerabilities to these challenges however are not uniform across regions of the world. Some continents are more vulnerable to the impacts of climate change and environmental degradation than others. For instance, it is feared that Africa might experience the most severe impacts of climate change than other parts of the world and it is the continent that is least prepared to handle these impacts [2,3].

Growing evidence has shown that Green House Gas emission such as Carbon Dioxide (CO_2) and Sulphur Dioxide (SO_2) are some of the major causes of climate change [4]. Even though Africa contributes relatively less to GHG in the world [4], the major determinants of GHG emission in this continent has not been sufficiently explored to give evidence based indices for effective mitigation and proactive adaptation policies implementation in the continent. The role of trade and economic activities on GHG emission and climate change mitigation drive also begs for scholastic inputs. In recent years, the agricultural sector had to face increased environmental challenges due to new production methods and intensified production systems adopted to meet continued population growth and new energy demands around the world [5]. Fossil energy sources such as oil and gas are extensively exploited in Africa but are mostly exported or wasted through leakages, or flaring [6] with attendant consequence of threat to increase in GHG and global warming. Nigeria is one of the leading producers and users of fossil fuel in the world but it appears that the oil resource could be a source of danger to her economy in the future if issues of sustainable environmental management (such as abatement of GHG emission and climate change adaptation strategies) are not taken seriously now. Unfortunately there are no sufficient data or studies to give empirical evidence about the actual major determinants of Green House Gas (GHG) emissions in sub-Sahara Africa (SSA). This situation, if unchanged can lead to the occurrence of the forecasts on the deleterious effects of climate change on livelihoods and economies predicted by IPCC [1] and Yohe et al. [7] more especially as there is sufficient evidence to show that CO_2 emission has significant relationship with global warming. Researchers [8] warned that the cost of limiting CO_2 emissions is likely to be much higher in developing countries such as Nigeria and other Sub-Sahara African countries due to their faster underlying growth rate. Even if they were allowed to double or triple their emissions over another 100 years, they may still face higher costs than developed countries under much more stringent targets, they noted. On the other hand, large reductions in man-made CO_2 emissions are possible on a global scale only if the developing countries also take action.

Given the foregoing background it is pertinent to conduct a study that is capable of uncovering some of the major drivers of GHG emissions in Africa, especially in its most populous country, Nigeria so as to obtain data for proactive policy making in combating and adapting to the problem of climate change in Africa; hence the need for this study.

1.1 Objectives of the Study

The research was designed to specifically ascertain the influences of commercial activities especially forestry trades, fossil fuel demand (petroleum) and agricultural production strategy (proxied by farm mechanization) on the level of GHG (CO_2) emissions in her economy. The study also identified other indirect factors which influence the level of GHG emission level in Nigeria and then discussed the implications of the findings for greener trade policy implementation and climate change adaptation/mitigation strategies in Nigeria and Africa.

1.2 Theoretical and Empirical Issues

It has been noted that the artificial inputs of energy (especially demand for fossil fuel energy), chemical products and agricultural labour or technological input necessary to maintain the agro-ecosystems and to reach the necessary production levels deeply alter natural biogeochemical cycles and can provoke serious damages to the environment: soil degradation; pollution of water; air and soil; loss of biodiversity and increased greenhouse gases emissions [9]. These are all examples of environmental damages which can be caused by agricultural practices, mining activities, industrial activities, technology application as well as other commercial ventures in the economy.

As the global population continues to grow, which is estimated to reach 9 billion by 2050, there is an increasing strain on the forest resources, agriculture and fisheries sectors to meet food security needs. This increases the quest for more "virgin" lands or forests. The recent upsurge in quest for forest lands by foreign investors in Africa has been associated with pollution and forestry depletion in parts of Africa with attendant dangerous consequences for the planet's ecosystem and trade. This has drawn the attention of many international organizations lately including Trade Policy Centre for Africa (TRAPCA) who made it a conference theme in November, 2011. Land acquisitions as well as arable crop land expansion can indirectly lead to negative environmental impacts as shown by Dossou [10] who observed the case of the municipalities neighbouring Cotonou, where rural emigrants who lost their lands settled en masse on "unserviced plots", leading to extreme pollution and health problems. Meanwhile, the conversion of forested and uncultivated lands is associated with biodiversity loss, degradation, diversion of water from environmental flows and loss of ecosystem services such as the maintenance of soil and water quality, as well as carbon sequestration [11,12]. Deforestation of tropical forests is reported to be contributing significantly to CO_2 emissions: estimates of carbon released range from *0.5* to *3* billion tons of carbon per year [13] relative to the *6* billion tons associated with current fossil-fuel use. Many observers argue that forest clearing is to a large extent uneconomic and mainly due to the absence of property rights for rain forests. If so, noted Nordhaus, a significant reduction of emissions might therefore be achieved at low economic cost through a cessation of forest clearing.

It has been noted [14] that deforestation cases are widespread in the context of increasing commercial pressures on land and deepening of forest depletion which is worsening global warming. Reports [15] indicated that deforestation in Nigeria is a major area of environmental concerns and indeed one of the most important issues of the last ten decades. The relationship between deforestation and GHG emissions was explained by Botkin and Keller [16] who noted that when forests are cleared and the trees are burnt or rot, carbon is released as carbon dioxide which then goes to increase the volume of greenhouse gas in the atmosphere that can combine with ozone in the ozone layer to deplete the protective layer of the atmosphere thus stepping up global warming.

Levels of Green House Gas (GHG) emission into the atmosphere (which includes Carbon Dioxide CO_2) levels have been associated with increase in climate change and hence much of the thinking to date on how to address climate change has focused on incrementally reducing GHG emissions – such as the commitment to reduce emissions to 5percent below 1990 levels under the Kyoto Protocol [17]. The United Nations Framework Convention on Climate Change (UNFCCC) identifies two responses to climate change: mitigation of climate change by reducing greenhouse-gas emissions and enhancing sinks and adaptation to the impacts of climate change. Most industrialized countries committed themselves, as signatories to the UNFCCC and the Kyoto Protocol, to adopting national policies and taking corresponding measures on the mitigation of climate change and to reducing their overall greenhouse-gas emissions [17,18]. The Kyoto Protocol recognizes a strong linkage between CO_2 emission reduction goals, emissions trading and the role of developing economies including sub-Sahara Africa [19].

Technology has also been reported to have some effects on level of GHG emissions. IPCC [20] noted that improvements in technologies and measures that can be adopted in three energy end-use sectors (commercial/residential/institutional buildings, transportation and industry), as well as in the energy supply sector and the agriculture, forestry and waste management sectors could drastically reduce the levels of greenhouse emissions globally. These incremental improvements are important first steps in addressing the global problem of climate change, which this paper attempts to address. Agricultural production systems and technology really have roles to play in reducing levels of GHG emissions. Greenhouse Gases emissions databases, Agri-Environmental statistics and indicators and environmental accounting frameworks are methodologies to monitor the environmental performance of the different countries [9]. FAO maintained that these tools have been recognized as useful for formulation of policies designed to provide an effective incentive structure for sustainable management of natural resources, ensuring that national agricultural practices are developed and implemented in a holistic approach. For instance IPCC [20] monetized the likely damage that would be caused by a doubling of CO_2 concentrations and noted that for developed countries, estimated damages were of the order of 1% of GDP. Developing countries including sub-Sahara Africa were expected to suffer larger percentage damages, so mean global losses of 1.5 to 3.5 percent of world GDP were therefore reported. IPCC [4] reported essentially the same range because more modest estimates of market damages were balanced by other factors such as higher non-market impacts and improved coverage of a wide range of uncertainties. Recently Stern [21] took account of a full range of both impacts and possible outcomes (i.e., it employed the basic economics of risk premiums) to suggest that the economic effects of unmitigated climate change could reduce welfare by an amount equivalent to a persistent average reduction in global per capita consumption of at least 5%. Including direct impacts on the environment and human health (i.e., 'non-market' impacts) increased their estimate of the total (average) cost of climate change to 11 percent GDP; including evidence which indicates that the climate system may be more responsive to GHG emissions than previously thought increased their estimates to 14 percent GDP. Using equity weights to reflect the expectation that a disproportionate share of the climate-change burden will fall on poor regions (which includes Sub-Sahara Africa) of the world increased their estimated reduction in equivalent consumption per head to 20 percent.

It has been established that over the past century human activities have been releasing GHGs at a rate unprecedented in geologic time. As a consequence of this acceleration in the rate of emissions, the concentration of GHGs in the atmosphere has increased by 30 percent, since pre-industrial times [22]. Examples of such anthropogenic activities include trade, agriculture, deforestation (or forestry activities), fossil energy or fuel consumption and those other activities associated with economic growth. Thus most structural models of climate–economy interactions have followed the Ramsey–Cass–Koopmans infinitely-lived agent framework [23,24,25]. According to González-Marrero, Lorenzo-Alegría and Marrero noted that the fact that the growth in the demand for transport in Spain over the last decade has exceeded that of GDP suggests that there must be other factors besides income to explain mobility and fuel consumption. Our present paper attempted to explore possible causes of fuel consumption which has been shown to be an exogenous variable in determination of GHG emission. Fossil fuel consumption has been, in large part, attributed to economic growth. According to Sharma [26], the Environmental Kuznets Curve (EKC) has been used to explain the relationship between the economic activities and the emission of pollutants and between the economic activity and the use of natural resources. The EKC hypothesis posits that environmental degradation initially exaggerates when a country's per capita income is low but over time, as the economy grows, environmental degradation falls. This results in an inverted U-shaped relationship between income and the use of natural resources and waste emissions. This branch of research on the relationship between economic growth and environmental pollution can be categorized into three strands [26]. This study intends to verify the claim that certain economic activities influence CO_2 emission levels.

Sharma reiterated that energy, such as crude oil, natural gas and coal, plays a major role in residential and industrial energy needs, transportation, and electricity. The burning of fossil fuel is essential in every country as it is used for the production of goods and services. While it is true that burning of fossil fuel emits a high amount of CO_2 and pollutes our environment, it has been empirically and theoretically shown that an increase in energy consumption results in greater economic activity [26]. It follows that higher economic growth (GDP) will have a positive effect on carbon dioxide emissions at least in the short-run. According to Hooi and Smyth [27] a boost in energy consumption results in higher GDP because, in addition to the undeviating effect of energy consumed for commercial use which stimulates higher rates of economic growth, higher energy consumption results in an increase in energy production. Thus, an increase in pollution emissions is expected due to fast economic growth and ensuing greater fossil fuel consumption. Our proposed model, because it is in growth form, is essentially a short-run model. Hence, a priori, we expect income to have a positive effect on emissions. Similarly, a higher consumption of energy, a pre-requisite for economic growth, will also lead to more emissions. Hence, a positive relationship between energy consumption and carbon emissions is expected. Trade is expected to have a positive effect on CO_2 emissions. This effect has roots in the Hecksher-Ohlin (H-O) trade theory [26]. This trade theory proposes that under free trade, developing countries (mostly middle and low income countries) would focus on the production of goods that are rigorous in factors in which they have a comparative advantage, such as labour and natural resources. Thus, trade causes the movement of goods produced in one country for either consumption or further processing. More consumption of goods and further processing of goods, which takes place due to greater trade openness, is a source of pollution. Hence, the H-O theory actually perceives that pollution is stimulated from further processing and manufacturing of goods, which results from greater trade openness [28].

1.3 Analytical Framework

In simultaneous equation models, unlike single equation models, what is a dependent (endogenous) variable in another equation appears as an explanatory (exogenous) variable in another equation [29,30]. Thus, there is a feedback relationship between the variables. This feedback creates the *simultaneity problem*, rendering OLS inappropriate to estimate the parameters of each equation individually. Besides, a simultaneous equation model may have *identification problem*. One of the several ways of resolving this problem is via the *order condition of identification*. An equation is said to be identified (has a unique statistical form) if it is exactly identified or over identified. If an equation is *exactly identified* or *overidentified*, it can only be estimated using Two Stage Least Squares (2SLS) or Zellner's [31] Seemingly Unrelated Regression (SURE) Model but not OLS. If it is overidentified, besides 2SLS, maximum likelihood methods can be used to estimate the coefficients. The system of equation that is similar to the foregoing can be exemplified as follows:

$$Y_{1t} = A_1 + A_2Y_{2t} + A_3X_{1t} + U_{1t}$$
$$Y_{2t} = B_1 + B_2Y_{1t} + B_3X_{2t} + U_{2t}$$

Where the Ys are the endogenous variables (e.g. CO_2 emissions in tons per annum and fossil fuel energy demand), the Xs the exogenous variables (such as agricultural land under cultivation, farm technology, trade as percentage of GDP, forestry income and growth rate of GDP); the As and Bs, respective intercepts and slope coefficients of the variables and the U's the stochastic error terms.

Jang and Koo [32] have used this model to identify the impact of weather variation on crop yield in the Northern Plains. Beasley [33] noted that SURE is relatively underutilized despite its robustness in analyzing multiple dependent variables. He therefore encouraged social researchers to make more use of it. According to this scientist, there many situations in educational and behavioral research in which multiple dependent variables are of interest. Oftentimes these variables may take the pattern of path analytic model, but there are many other cases where they do not. However, it is commonplace for educational researchers to conduct separate analyses for multiple dependent variables even though they are likely to be correlated and have similar although not identical design matrices. For example, researchers in counseling often have multiple outcomes (measure of symptoms, coping, etc.) that are assumed to have some of the same predictors but to also have predictors that are unique to each measure. This is a situation that calls for a SUR model; however, a search of ERIC and PSYCHINFO located 11 applications of SUR models despite the enormous number of articles that analyze multiple dependent variables [33].

2. RESEARCH METHODS

2.1 Study Area

Nigeria is in West African sub-region; bordering the North Atlantic Ocean, between Benin Republic and Cameroon. Nigeria has a total land area of 923,773 square kilometers populated by over 140,003,542 people (going by 2006 population census). Her major revenue earner is crude oil. Climate varies - equatorial in south, tropical in centre, arid in north. Average rainfall hovers around 1282.2 mm varying from 500 - 1800mm. In 2007 agriculture contributed 42.08 percent to Nigerian's GDP. Out of this figure, crops, livestock, forestry and fishing contributed 37.54 percent, 2.64 percent, 0.53 percent and 1.37 percent to the country's economy respectively. Agricultural *Products-* include cocoa, palm oil, yams, cassava, sorghum, millet, corn, rice, livestock, groundnuts, cotton. Industry types include textiles, cement, food products, footwear, metal products, lumber, beer, detergents and car assembly [34].

2.2 Sampling and Data Collection Method

Secondary data, mainly time series data from Central Bank of Nigeria's Annual Report, World Bank data and Nigerian Bureau of Statistics data were used for this study. The data collected covered a period of 40 years (1970 – 2009).

2.3 Data Analysis Method

The data gathered were analyzed using Zellner's Seemingly Unrelated Regression model. The specific model used is given as follows:

Equation 1 $CO_2 = \alpha o + \alpha_1 \text{ agriclanda} + \alpha_2 \text{ farmtech} + \alpha_3 \text{ forest} + \alpha_4 \text{fsflendemand} + \alpha_5 \text{tradeperctgdp} + u_1$

Equation 2 $\text{fsflendemand} = \beta_0 + \beta_1 \text{ farmtech} + \beta_2 \text{ tradeperctgdp} + \beta_3 \text{gdpgrwthrate} + u_2$

Where the CO_2 = level of CO_2 emissions in kilo tons per annum; fsflendemand = fossil fuel energy demand in millions of naira per annum. The exogenous variables include: agriclanda = agricultural land under cultivation (in thousands of hectares per year), farmtech = farm technology proxied by number of tractors/farm machineries in thousands acquired per year, tradeperctgdp = aggregate trade in the economy per annum as percentage of GDP (in percentage), forestry income (in millions of naira) and growth rate of GDP in percentage; the As and Bs, respective intercepts and slope coefficients of the variables and the U's the stochastic error terms.

3. RESULTS AND DISCUSSIONS

Results of the model estimates are presented in Table 1. The model fitness test is presented at the top most side of the table. From the table it would be seen that the primary equation, CO_2 recorded an R^2 of 0.70, implying that 70 percent variation in the CO_2 emission levels in Nigerian economy estimated in this model was explained by the variables in the first equation. This indicates a very good fitting. The Chi-square estimate which tests the null hypothesis of no joint effects of the independent variables of the model was significant at 1 percent statistical level. This further buttresses the fitness of the model. An evaluation of the second equation shows that the fossil energy fuel demand equation indicated an R^2 of about 0.50, implying that about half of the increase in level of fossil energy or fuel demand (which in turn could exert some influences on the level of CO_2 emission) was explained by the exogenous variables included in the second equation. The results of the Breusch-Pagan test (test for serial correlation) which gave a Chi-square estimate of 0.004 at p=0.95, indicates that there is no dependency in the errors of the two equations gave us the room to accept the null hypothesis of no interdependence of errors and conclude that our model is free of such dependence.

From the primary equation, i.e. the CO_2 equation, it is indicated that all the explanatory variables of the model, except forestry activities (proxied by forestry income) exerted positive influences on the level of CO_2 emission in Nigerian economy. This implies that increase in any of these variables, i.e. agricultural land under cultivation, farm technology, trade as percentage of GDP and fossil fuel energy utilization or demand is associated with an increase in the level of CO_2 emissions in the nations' environment. Interestingly, almost all the hypothesized factors returned statistically significant slope coefficients except trade as a percentage of GDP.

Table 1. Seemingly unrelated regression results

Equation	Obs	Par	RMSE	R-Sq	Chi2	P
CO_2	40	5	11516.34	0.70	94.95	0.000
Fsflen demand	40	3	2.890085	0.49	38.51	0.000
			Coef.	**Std.**	**z**	**P>|z|**
CO_2						
Agriclanda			0.38	0.13	2.92***	0.00
Farmtech			4590.51	2470.54	1.86*	0.06
Forest			-9509.46	2423.59	-3.92***	0.00
fsflendemand			1542.99	578.77	2.67***	0.01
Tradeperctgdp			-219.96	181.23	-1.21NS	0.23
Intercept			-226080.00	90486.65	-2.50***	0.01
Fsflendemand						
Farmtech			1.226	0.450	2.730***	0.006
Tradeperctgdp			0.000	0.039	0.000NS	1.000
gdpgrwthrate			-0.260	0.075	-3.470***	0.001
Intercept			11.573	1.351	8.570***	0.000
Correlation matrix of residuals						
	CO_2	fsflendemand				
	CO_2	1				
Fsflendemand	0.0103	1				
Breusch-Pagan	test of	independence:	chi2(1)	=	0.004, Pr	= 0.948

Source: Analysis of CBN and World Bank Data using STATA by Authors (2012). NB: (***) = Figures significant at p =<0.01, (**) = Figures significant at p =<0.05, (*) = Figures significant at p =<0.10, while NS = Not significant below p =0.10.

Agricultural land area under cultivation (agriclanda) returned a Z-value of 2.92, which was statistically significant at p<0.01. This shows that the probability of this factor increasing the levels of CO_2 emission in the country's environment is very significant and not by chance. The finding is in line with FAO's [9] worry that expansion of forest lands for agricultural expansion and even the rush for land by foreigners to invest in agriculture [14] will not do Africa any good rather it will worsen the environmental problem, particularly the problem of increased CO_2 emission in the continent's environment. This fear is even more pronounced when one observes that accompanying farm technology (especially tractorization and use of farm machineries) as seen in our model parameter estimate (Table 1) also indicated that farm technology adoption is associated with increase in the level of CO_2 emissions in Nigeria. This variable is statistically significant at p<0.10. The findings justifies the fears of all those who are campaigning against land grabbing in Africa. Contrary to our expectations on forestry activities, however, the forestry income variable which represent the level of forestry activities in Nigeria by our model did not show a positive sign but instead returned a negative sign which is statistically significant at p<0.01 with a Z-statistic of -3.92. This may be construed to be a sign that forestry activities or explorations are still being carried out sustainably in Nigeria at the period in review. However, after some years of continuous forestry exploitation, a threshold will be reached when the activities of forestry such as timber exports and utilization of fuel wood will combine to bring about significant forest cover loss and carbon sequestration drive maybe jeopardized. It would be recalled that Nordhaus [11] indicated that forestry activities or deforestation has a significant impact on GHG emission. This assertion is related to the next findings which indicated that fossil fuel energy demand in Nigerian economy within the period in review by this study is a very significant determinant of CO_2 emission in the country with an estimated Z-statistic which was found significant at $p<0.01$. This finding has equally vindicated FAO [9] who earlier noted that energy inputs (especially fossil fuel utilization) in the economies of many nations are partly and largely responsible for pollution or CO_2 emissions in developing and developed countries as well. There are policy implications for this which will be discussed later in our conclusion. The second equation's parameter estimates justifies the significance of including fossil fuel energy demand in the economy as a major variable in the emission of GHG determination as well its classification as an endogenous variable. As we earlier noted, the exogenous variables in this equation exhibited a fairly high coefficient of variation and in addition two out of its explanatory variables were found to be significant determinants of fossil fuel energy demand, thus enabling these factors to be regarded as indirect contributors to the CO_2 emission level in the nation's environment. It would be observed that through this factor for instance, economic growth rate (which is a product of all commercial activities growth, exports and imports of both oil and non-oil products, mining and manufacturing) indicated a significant effect on the level of fossil fuel demand in the economy over the period in review. This variable has an estimated Z-statistic of 3.470 and is statistically significant at $p<0.01$. The findings affirms the fears of environmental scientists [16] and other institutions/stakeholders such as World Bank [2], IPCC [4,17,14] who expressed worries over the possible effects of economic growth on energy demands and global warming.

4. CONCLUSION

Through the chosen econometric approach of this study it has been shown that agricultural land area expansion in various forms, be it through land grabbing by foreign investors or by internal policies of the nation's agricultural policy, along side other factors such as farm technology based on increasing use of farm machineries and tractors; deforestation (forestry commercial activities) and fossil fuel energy demand all contribute significantly towards the level of GHG emission (CO_2) levels in Nigerian environment. Besides these factors there are other factors which influence pollution or GHG emissions indirectly. These enter the system through the effects of economic activities that promote GDP growth rate as well as the chosen technology for agricultural production (given that agriculture is a major employer of Nigerian burgeoning population) via the use of fossil fuel energy. The demand for fossil energy thus appears to be one of the most significant issues to tackle if the problem of climate change mitigation in Nigeria has to be given the seriousness it deserves. In light of the foregoing findings we recommend that Nigeria should invest in energy efficient technologies and should utilize less of fossil fuels. Agricultural land expansion programmes and land grabbing should be moderated by the governments to check excessive opening and depletion of forestry resources in Nigeria. Since agriculture engage more than 65 percent of Nigerians, efforts should be made by the various governments at different levels to assist farmers adopt climate resilient technologies which will also ensure sustainable agricultural production. They need to be encouraged to adapt to the looming dangers of climate change now. Nigeria should put in place policies that will tax companies or firms emitting GHGs and utilize such tax proceeds for research and building the capacities of farmers to adapt to deleterious effect of climate change in the country and continent. The development of existing and new technologies for adapting to climate change and variability, building of environmental consciousness of Nigerians through curriculum restructuring and provision of weather information services by the governments to enable farmers plan against weather uncertainty and risks are hereby recommended.

COMPETING INTERESTS

Authors have declared that no competing interests exist.

REFERENCES

1. IPCC Climate change 2007: impacts, adaptation and vulnerability. Contribution of Working Group II to the Fourth Assessment Report of the Intergovernmental Panel on Climate Change; Cambridge University Press, Cambridge, United Kingdom and New York, NY, USA; 2007.
2. World Bank The costs to developing countries of adapting to climate change new methods and estimate the global report of the economics of adaptation to climate change study consultation draft. Washington D. C.: The World Bank Group; 2010.
3. TerrAfrica. Land & climate: The role of sustainable land management (SLM) for climate change adaptation and mitigation in Sub-Saharan Africa (SSA); 2009.
4. IPCC, Climate Change 2001: The Scientific Basis. Contribution of Working Group I to the Third Assessment Report of the Intergovernmental Panel on Climate Change, Houghton JT, Ding Y, Griggs DJ, Noguer M, van der Linden PJ, Dai X, Maskell K. and Johnson CA, Eds., CambridgeUniversity Press, Cambridge. 2001;881.
5. Seo SN, Mendelsohn R, Dinar A, Hassan R, Kurukulasuriya. A Ricardian analysis of the distribution of climate change impacts on agriculture across agro-ecological zones in Africa. Environmental and Resource Economics. 2009;43:313-332.
6. OECD Development Centre/African Development Bank. Growth trends and outlook for Africa: Time to unleash Africa's huge energy potential against poverty concludes. African Economic Outlook; 2003-2004.
7. Yohe GW, Lasco RD, Ahmad QK, Arnell NW, Cohen SJ, Hope C, Janetos AC, Perez RT. Perspectives on climate change and sustainability. Climate change 2007: Impacts, adaptation and vulnerability. contribution of Working Group II to the Fourth Assessment Report of the Intergovernmental Panel on Climate Change, Parry ML, Canziani OF, Palutikof JP, van der Linden, PJ and Hanson CE. Eds., Cambridge University Press, Cambridge, UK. 2007;811-841
8. Hoeller P, Dean A, Nicolaisen J. Macroeconomic implications of reducing greenhouse gas emissions: a survey of empirical studies. OECD Economic Studies;1991. Retrieved on 17th July, 2012.
    Available: http://www.oecd.org/dataoecd/47/58/34281995.pdf
9. FAO monitoring the interaction between agriculture and the environment: current status and future directions African commission on agricultural statistics 22nd session Addis Ababa, Ethiopia, 30 November - 3 December; 2011. Retrieved on 23/6/2012.
    Available: http://faostat.fao.org
10. Dossou PJ. Evolution and impacts of coastal land use in Benin: The case of the Sèmè-Podji commune. VADID contribution to ILC Collaborative Research Project on Commercial Pressures on Land, Rome; 2011.
11. African Union Framework and guidelines on land policy in Africa: a framework to strengthen land rights, enhance productivity and secure livelihoods. African Development Bank/African Union/Economic Commission for Africa; 2009.
12. Markelova H, Meinzen-Dick R. The importance of property rights in climate change mitigation. 2020 Vision Briefs 16(10), International Food Policy Research Institute (IFPRI); 2009.

13. Nordhaus WD. To slow or not to slow: the economics of the greenhouse effect, revision of a paper presented to the 1989 meetings of the International Energy Workshop and the MIT Symposium on Environment and Energy; 1990.

14. Anseeuw WL. Alden W, Cotula L, Taylor M. Land rights and the rush for land: findings of the global commercial pressures on land research project. ILC, Rome; 2009. Retrieved on 20th May, 2012. Available: http://www.landcoalition.org/cplstudies

15. Onoja AO, Idoko C, Adah C. Policy implications of the effects of deforestation on Nigerian economy. Production and Agricultural Technology Journal, PAT. 2008;4(2):114-120. Retrieved on 26th September, 2008.
Available: http://www.patnsukjournal.com/currentissue

16. Botkin DB, Keller EA. Environmental science: earth as a living planet. 2nd Ed. New York: John Wiley Press. 1997;215.

17. UNDP. Human Development Report 2007/2008. Fighting climate change: Human solidarity in a divided world, Palgrave Macmillan; 2007-2008.

18. Klein RJT, Huq F, Denton TE, Downing RG, Richels JB, Robinson, Toth FL. Inter-relationships between adaptation and mitigation. Climate Change 2007: Impacts, Adaptation and Vulnerability. Contribution of Working Group II to the Fourth Assessment Report of the Intergovernmental Panel on Climate Change, Parry ML, Canziani OF, Palutikof JP, van der Linden PJ, Hanson CE. Eds. Cambridge University Press, Cambridge, UK. 2007;745-777.

19. Ellerman AD, Jacoby HD; Decaux. The effects on developing countries of the kyoto protocol and CO_2 emissions trading. Joint Program on the Science and Policy of Global Change Massachusetts Institute of Technology. (n.d.); 1998.

20. IPCC Impacts, adaptations and mitigation of climate change: Scientific-technical analyses. Contribution of Working Group II to the IPCC; 1996.

21. Stern N. *The economics of climate change: the Stern Review*. Cambridge University Press, Cambridge. 2007;692.

22. Antuasegi A,Escapa M. Economic growth and greenhouse gas emissions. *Ecological Economics*. 2002;40:23–37. Retrieved on 17th July, 2012.
Available: http://upi-yptk.ac.id/Ekonomi/Ansuategi_Economic.pdf

23. Ramsey F. A mathematical theory of saving. Econ. J. 1928;38:543–549.

24. Cass D. Optimum growth in an aggregative model of capital accumulation. Rev. Econ. Stud.1965:32:233–240.

25. Koopmans TC. On the concept of optimal economic growth. In: The Econometric Approach to Development Planning. North Holland, Amsterdam; 1965.

26. Sharma SS. Determinants of carbon dioxide emissions: Empirical evidence from 69 countries. Applied Energy. 2011;88:376–382

27. Hooi L, Smyth R. CO_2 emissions, electricity consumption and output in ASEAN. Applied Energy. 2010;87:1858–64.

28. Halicioglu F. An econometric study of CO_2 emissions, energy consumption, income and foreign trade in Turkey. Energy Policy. 2009:37:699–702.

29. Gujarati DN. Essentials of econometrics 3rd Edition. New York: Mc-Graw Hill; 2006.

30. Koutsoyiannis A. Theory of Econometrics: An Introductory Exposition to Econometric Methods, MacMillan N.Y.; 1981.

31. Zellner A. An Efficient Method of Estimating Seemingly Unrelated Regression Equations and Tests of Aggregation Bias, Journal of the American Statistical Association. 1967;57:500-509.

32. Jang Y, Koo W. Identifying the Impact of Weather Variation on Crop Yield in the Northern Plains. Poster prepared for presentation at the Agricultural & Applied Economics Association 2011 AAEA & NAREA Joint Annual Meeting, Pittsburgh; 2011. Pennsylvania, July 24-26.Retrieved on 17th July, 2012.
     Available:http://ageconsearch.umn.edu/bitstream/104508/4/SelectedPoster_WeatherI mpactOnYield.pdf

33. Beasley TM. Seemingly Unrelated Regression (SUR) Models as a solution to path analytic models with correlated errors. Multiple Linear Regression Viewpoints. 2008;34(1). Retrieved on 17th July, 2012.
     Available: http://mlrv.ua.edu/2008/vol34_1/Beasley-SUR.pdf

34. Central Bank of Nigeria. Annual Report and Statistical Bulletin. Central bank of Nigeria: Abuja; 2007.

Permissions

List of Contributors

Sunil Baidar
Department of Chemistry and Biochemistry, University of Colorado, Boulder, CO, USA
Cooperative Institute for Research in Environmental Sciences, Boulder, CO, USA

Rainer Volkamer
Department of Chemistry and Biochemistry, University of Colorado, Boulder, CO, USA
Cooperative Institute for Research in Environmental Sciences, Boulder, CO, USA

Raul Alvarez
Earth System Research Laboratory, NOAA, Boulder, CO, USA

Alan Brewer
Earth System Research Laboratory, NOAA, Boulder, CO, USA

Fay Davies
School of Built Environment, University of Salford, Salford, UK

Andy Langford
Earth System Research Laboratory, NOAA, Boulder, CO, USA

Hilke Oetjen
Department of Chemistry and Biochemistry, University of Colorado, Boulder, CO, USA

Guy Pearson
Halo Photonics, Worcestershire, UK

Christoph Senff
Cooperative Institute for Research in Environmental Sciences, Boulder, CO, USA
Earth System Research Laboratory, NOAA, Boulder, CO, USA

R. Michael Hardesty
Cooperative Institute for Research in Environmental Sciences, Boulder, CO, USA
Earth System Research Laboratory, NOAA, Boulder, CO, USA

William S. Sicke
Department of Civil and Environmental Engineering, University of California, Davis, USA

Jay R. Lund
Department of Civil and Environmental Engineering, University of California, Davis, USA

Josué Medellín-Azuara
Department of Civil and Environmental Engineering, University of California, Davis, USA

Yu-De Huang
Environmental Engineering Research Center, Sinotech Engineering Consultants, Inc., 6F, No.280, Xin Hu 2nd Rd., Nei Hu Dist, Taipei City, Taiwan

Hsin-Hsu Huang
Environmental Engineering Research Center, Sinotech Engineering Consultants, Inc., 6F, No.280, Xin Hu 2nd Rd., Nei Hu Dist, Taipei City, Taiwan

Ching-Ping Chu
Environmental Engineering Research Center, Sinotech Engineering Consultants, Inc., 6F, No.280, Xin Hu 2nd Rd., Nei Hu Dist, Taipei City, Taiwan

Yu-Jen Chung
Environmental Engineering Research Center, Sinotech Engineering Consultants, Inc., 6F, No.280, Xin Hu 2nd Rd., Nei Hu Dist, Taipei City, Taiwan

Jonathan Muñoz
NOAA-Cooperative Remote Sensing Science and Technology Center (NOAA-CREST), City College of New York, 160 Convent Ave, NY 10031, USA

Jose Infante
NOAA-Cooperative Remote Sensing Science and Technology Center (NOAA-CREST), City College of New York, 160 Convent Ave, NY 10031, USA

Tarendra Lakhankar
NOAA-Cooperative Remote Sensing Science and Technology Center (NOAA-CREST), City College of New York, 160 Convent Ave, NY 10031, USA

Reza Khanbilvardi
NOAA-Cooperative Remote Sensing Science and Technology Center (NOAA-CREST), City College of New York, 160 Convent Ave, NY 10031, USA

Peter Romanov
NOAA-Cooperative Remote Sensing Science and Technology Center (NOAA-CREST), City College of New York, 160 Convent Ave, NY 10031, USA

Nir Krakauer
NOAA-Cooperative Remote Sensing Science and Technology Center (NOAA-CREST), City College of New York, 160 Convent Ave, NY 10031, USA

Al Powell
NOAA/NESDIS/Center for Satellite Applications and Research (STAR) 5200 Auth Road, WWB, Camp Springs, MD 20746, USA

T. Mubyana-John
Department of Biological Sciences, University of Botswana, P/Bag 0022, Gaborone, Botswana

W. R. L. Masamba
Okavango Research Institute, University of Botswana, P/Bag 285, Maun, Botswana

Mohammad Tarikul Islam
Department of Government and Politics, Jahangirnagar University, Bangladesh

A. I. Achike
Department of Agricultural Economics, University of Nigeria, Nsukka, Nigeria

Anthony O. Onoja
Department of Agricultural Economics and Extension, University of Port Harcourt, Nigeria

Printed in the USA
CPSIA information can be obtained
at www.ICGtesting.com
JSHW051447221024
72173JS00006B/1604

9 781682 860014